ON THE EVOLUTION
OF HUMAN BEHAVIOR

ON THE EVOLUTION OF HUMAN BEHAVIOR

THE ARGUMENT FROM ANIMALS TO MAN

PETER C. REYNOLDS

UNIVERSITY OF CALIFORNIA PRESS
Berkeley, Los Angeles, London

University of California Press
Berkeley and Los Angeles, California
University of California Press, Ltd.
London, England

Library of Congress Cataloging in Publication Data

Reynolds, Peter C
  On the evolution of human behavior.

  Includes index.
  1. Genetic psychology. 2. Animals, Habits and
behavior of. 3. Human evolution. I. Title.
BF701.R39   1981   155.7   80–6056

Printed in the United States of America

I have already explained to you what is out of the common is usually a guide rather than a hindrance. In solving a problem of this sort, the grand thing is to be able to reason backward. That is a very useful accomplishment, and a very easy one, but people do not practise it much. In the everyday affairs of life it is more useful to reason forward, and so the other comes to be neglected.

SHERLOCK HOLMES,
in *A Study in Scarlet*
by SIR ARTHUR CONAN DOYLE

The page is largely blank with only a faint, illegible block of text near the center that cannot be reliably read.

CONTENTS

PREFACE AND
ACKNOWLEDGMENTS

A book cannot be said to have an exact beginning in time, so I cannot mention all of the people to whom I am indebted professionally and personally. I must single out those whose pursuits have influenced the direction of the present work or whose assistance helped bring it to completion. Professor Karl Pribram provided continuing enthusiasm and support for my inquiry into the argument from animals to man, and I cannot imagine the present formulation without his part in it.

There are also a number of other people from whom I learned a great deal: Bill Abler, the Sensuous Gadgeteer, who emphasized that imagination is terribly important in even the simplest constructional tasks, and Barbara Honegger who taught me facts about verbal memory I know I could not have learned anywhere else. The late J. Bronowski foresaw a biology of language, and through his kindness I was able to witness the extension of linguistics to sign language by Ursula Bellugi and Edward Klima. Mention should also be made of Mike and Evelyn, who first clearly demonstrated to me the human capacity for internal imagery, and Jane Lancaster who gave me my first contact with primate social behavior. I cannot think of phonemes without Alvin Liberman's explication of them, and James Dewson's experimental finesse enhanced my understanding of science. Richard Henderson emphasized that knowledge of an idea's history can save you many errors. Without Adam Kendon, I might never have learned to watch movies one frame at a time, and Tim and Patsy Asch enhanced my appreciation of film as both art and science. I am also grateful to Kirk and Karen Endicott for introducing me to the Batek people of Malaysia, from whom I learned that simple technologies are never simple.

Derek Freeman founded the Laboratory of Human Ethology at the Australian National University, and Roger Keesing helped make it possible to write this book under its auspices. Without James Fox's invitation, I might never have done so. Henny Fokker's bibliographic talents were invaluable, and Ann Buller has helped in many matters great and small.

Mark Noble and Colin Groves each kindly read the manuscript with the eye of both author and biologist. Moreover, I owe special thanks to Harold C. Conklin, whose broad conception of linguistic anthropology made possible my peregrinations, and to Hilary Yerbury, whose knowledge of French language and history contributed to my own. Finally, a complete account requires that I mention those companions who never read a word but whose intellectual influence was decisive: the three dozen rhesus monkeys I came to know well.

Canberra, Australia
April 1979

INTRODUCTION

Many books have been written in the past few years about human evolution, behavioral biology, primate social organization, and animal communication. There has also been much debate on such issues as sociobiology and the nature of culture. Unfortunately, the debate about behavioral biology and ethology has often been conducted in terms of oppositions and categories invalidated by the very advances in knowledge that have animated these discussions. Many of these debates seem to regard ethology as just one more form of biological determinism that will eventually go away and leave social scientists in peace. Ethologists themselves have contributed to this impression by applying a theory of instinct devised for birds and fish to large-scale interactions among human institutions, without any attempt to deal with the question of why humans have institutions and what this fact may hold for their theory of instinct. Nonetheless, ethology will not go away because it challenges the assumptions of the Western intellectual tradition prevailing for the last two centuries. Whether ethologists themselves are aware of this situation is immaterial for assessing the implications of ethology for the history of thought. Because the assumptions in question are also the founding axioms of most modern social science, ethology cannot be accommodated by bringing it within the gates. It is a Trojan horse, and its effect is necessarily a radical transformation of the theory and practice of normal social science. An anthropological interpretation of ethology is not a list of the instincts of the human species. It is an assessment of its implications for our understanding of man and society. Whatever the ultimate effects of ethology, one conclusion is clear: the self-consistent framework for the human sciences constructed by the eighteenth century is no longer intellectually credible. With the coming of age of ethology, the Enlightenment has come to an end.

LORD MONBODDO'S TAIL

DISCOURSE ON THE ORIGINS OF THE ORIGIN OF INEQUALITY

Is social inequality ordained by natural law? This was the subject of an essay contest sponsored by the Academy of Dijon in France in 1753, and it was answered by Rousseau in his first major work, *The Discourse on the Origin of Inequality Among Men*, completed early the following year.[1] This essay has been both widely praised as a vindication of the simple life and roundly dismissed as a romanticized account of a fanciful state of nature. Adam Smith reviewed it favorably for the *Edinburgh Review;* but Boswell, in September 1769, when longing for primitive contentment, was admonished by Dr. Johnson with the observation that "Rousseau *knows* he is talking nonsense and laughs at the world for staring at him."[2] Although Lévi-Strauss (1962) has called Rousseau one of the founders of the sciences of man, many historians of anthropology, if they mention Rousseau at all, still regard him as the apostle of the noble savage.[3] It was about the *Discourse on Inequality* that Rousseau later complained, "Nowhere in Europe did it find more than a few readers who understood it, and not one of them chose to speak of it."[4] Yet it was Rousseau, in this essay, who was among the first to suggest the use of nonhuman primates to shed light on the social evolution of man, and he was the first to give social anthropology the mandate for its work.[5]

Although the intellectual context has changed dramatically in the past two centuries, the issues raised by Rousseau continue to be germane to anthropology. There are people today who still seem to be haunted by the relationship between social inequality and natural law, and the question cannot be dismissed as having been superseded by the progress of anthropological science. There has been progress in our understanding of man, but it is not a theoretical progression, analogous to idealized history of physical science, where a less general theory is replaced by the more general, as Einstein's physics replaced that of Newton. We can no longer ask how interstellar ether affects the speed of light, yet the question of the relationship between human nature and social hierarchy has been not invalidated but only precluded by the assumptions of modern social science.

Anthropology, to be understood as the natural history of human behavior and custom, has made great methodological progress in the two centuries since Rousseau. However, in the conventional anthropological history, the development of this methodological sophistication has been portrayed as a theoretical progression, in which the "conjectural history" of the Enlightenment gave way to the "evolutionism" of the Victorians and, finally, to the triumph of "structural-functionalism" early in the present century. What is most damaging about such a historical perspective is that it portrays the development of anthropology as the replacement of one banal theory by another; and in doing so, it obscures the real achievement of anthropological science, which is not so much a matter of intellect as of attitude. What anthropology has accomplished, and here it is very much a child of the Enlightenment, is the institutionalization of a moral injunction: the willingness to look at other societies as if they were created by human beings with as much intelligence and integrity as ourselves. For this development, Piaget's term *decentration* is most appropriate.[6] The child who sees the world as organized in relation to himself and following his own concerns exhibits a lack of decentration. *Objectivity,* a related term often used in this context, is too optimistic, for it implies a correspondence between our mental models and the world of facts, whereas decentration implies only the preconditions to objectivity while accepting the possibility of error.

If the history of anthropology has been primarily the history of decentration, and not the history of wrong ideas superseded by correct ideas, then the sin of early anthropologists was perhaps not conceptual at all: perhaps not an error but an actual sin; not a mistake about facts but a deficiency of character. Like Lowie before us (1937, p. 24), we need only cite the remarks of one of the eminent anthropologists of the Victorian age, Sir John Lubbock (1869, pp. 557–558), friend of the Darwin family, whose opinions figure prominently in *The Descent of Man:*

> The harsh, not to say cruel, treatment of women, which is almost universal among savages, is one of the deepest stains upon their character. They regard the weaker sex as beings of an inferior order, as mere domestic drudges. Nor are the labours and sufferings of the women sweetened by any great affection in the part of those for whom they work.[7]

FIGURE 1.1   From *Sex and the Social Order* by Georgene H. Sew-ard. Copyright 1946, McGraw-Hill. *(Used with the permission of McGraw-Hill Book Company.)*

It we compare these words to the scene shown in figure 1.1, illustrating an aspect of Victorian England Lubbock must have known, how else are we to interpret such a juxtaposition except to suppose that the Victorians thought not in terms of a science of man but in terms of a science of savages that had few points in common with a science of Europeans?

Nineteenth century anthropological theory was an inadequate basis for a science of man, not because it postulated evolutionary stages of culture, but because it was predicated on a Eurocentric view of history which made it impossible to understand non-Western societies on their own terms. Yet modern textbooks of anthropology, by the characterization of Victorian ethnologists as "unilinear evolutionists" have obscured the most important differences between the Victorian's understanding of evolution and our own. The anthropologists of the last century thought of themselves as empiricists, and they provided impressive empirical confirmation of an evolutionary approach to human history. They established the existence of an extinct species of fossil man, demonstrated the coexistence of Stone Age man with extinct species of animals, discovered the existence of glacial ages in the human past, catalogued the succession of the major Mediterranean cultures, and established the genealogies of the world's major languages.[8] Nonetheless, this anthropology did not have an evolutionary theory of society in a modern sense of evolution and it did not think of history in a way that would be acceptable to a modern historian.

The nineteenth century developed the methods of historical scholarship and archeological excavation, but the larger intellectual context was still dominated by a conception of man and society that was not the product of these empirical investigations but its inheritance from the previous century. The inadequacy of Victorian anthropology was not that its evolutionism was unilinear but that it was not yet truly evolutionist. Evolution, as we now understand that term, is preeminently a historical concept that recognizes the great possibilities for change in both the products of nature and the works of man, and it sees these changes as part of a historically continuous chain of events. It is a dynamic conception of nature and history most explicitly developed in the nineteenth century by Darwin and by Marx. The evolutionism of Victorian anthropology, in contrast, was a typological framework which did not emphasize the process of historical change but the ranking of enduring forms on a scale of temporal progression.

The distinction between Darwinian and typological evolutionism is a subtle one, and it is not a distinction that could easily be made by nineteenth-century thinkers themselves. The implications of Darwinian evolution have only become apparent in our own time, and it is clear that Darwin's thought was a more original departure from traditional evolutionism than was generally appreciated by his contemporaries. Although seemingly one with the general atmosphere of nineteenth-century thought, Darwin's theory was founded upon a radically different system of nature than that which sustained the biological evolutionism of his predecessor Lamarck or the social evolutionism of his friend Sir John Lubbock. Also, it was a major departure from the well-regulated machine of Newtonian science.

The failure to recognize the novelty of Darwin is best exemplified by Samuel Butler, the author of *Erewhon,* in a critique of Darwin written in 1879. He objected to the mechanism of natural selection, which he correctly saw as Darwin's original contribution; but the idea of the evolution of species he traced back to biologists of the previous century, Lamarck, Erasmus Darwin, and Buffon. In Butler's interpretation, Darwin's transformist hypothesis was in conformity with a long line of evolutionary theory, and it differed only in its idiosyncratic adherence to the mechanism of natural selection. Revealing, too, is Butler's contention that Count Buffon was an evolutionist, for this last-mentioned author was an outspoken critic of the evolution of species, an idea that had been put forth in a vague and conjectural form by his colleague Maupertuis, in the 1750s.[9] Yet in Buffon's encyclopedic *Natural History,* one of the most widely read books of its age,[10] Butler came upon the following passage and used it as the basis of his argument:

> The body of a horse, for example, which by a single glance of the eye, appears so different from the body of a man, when it is compared part by part, instead of surprising by the difference, only

astonishes by the singular and almost perfect resemblance. In fact, take the skeleton of a man, bend downwards the bones of the pelvis, shorten those of the thighs, legs, and arms, lengthen those of the feet and hands, join the phalanges, lengthen the jaws, by shortening the frontal bone, and extend the spine of the back, this skeleton would cease to represent the remains of a human figure, and would be the skeleton of a horse: for it is easy to suppose that in lengthening the spine of the back and jaws we augment, at the same time, the number of vertebrae, ribs and teeth; and it is only by the number of those bones, which may be looked upon as accessory, and, by the prolongation, the shortening, or junction, of the others, that the skeleton of a horse differs from that of the human body. We see in the description of the horse these facts too well established to doubt; but, to follow these relations still farther, let us consider separately some essential parts of the structure; for example, we find ribs in all quadrupeds, in birds, and in fish; and we find the vestiges even in the shell of the turtle. Let us also consider, that the foot of a horse, so different in appearance from the hand of a man, is, notwithstanding, composed of the same bones, and that we have, at the extremity of each of our fingers, the same little bone resembling a horseshoe, which terminates the foot of that animal.[11]

Historians of science have questioned the interpretation that Buffon was an evolutionist, and their argument is supported by the fact that later anatomists, who had actual proof of the extinction of species, were not evolutionists either (Fellows and Milliken 1972; Wilkie 1959). Evolution is above all a historical concept. What Buffon was expressing in his article on the ass was a concept that is only superficially historical: the concept of homology, or the formal correspondence of organs of different species. To interpret such topological correspondences among animals as derivative of their actual history is to draw a conclusion from the data — no matter how obvious it is to us — that most men of the eighteenth century manifestly did not draw. Intellectually, Buffon and Darwin lived in different worlds, and Darwin's reinterpretation of the concept of homology is a measure of the distance between them.

The concept of homology (Ghiselin 1976), the mapping of similar parts across species, was a method already in use in the comparative anatomy of the seventeenth century, and it was widely adopted, not only by other natural sciences, but by the nascent science of man.[12] It came to prominence, however, in an intellectual context that was still dominated by what Lovejoy has called "the great chain of being" (1936). Though the concept and its expression is a continuing one in Western thought, traceable to Plato in the *Timaeus*, the great chain was given contemporary expression in the eighteenth century by the progenitors of modern biology and social science. So, too, in the philosophical background to that century, thinkers as dissimilar in their way as Leibniz was from Locke were nonetheless agreed on the *scala naturae*.[13] John Locke, that inspiration to the Enlightenment, cited by Rousseau in the *Discourse on Inequality* and compared by D'Alembert to Newton in the first volume of the *Encyclopedia*,[14] expressed it thus in *Essay Concerning Human Understanding* (1975, pt. 3, chap. 6, par. 12, Variorum ed.):

> In all the visible corporeal world we see no chasms or gaps. All quite down from us the descent is by easy steps, and a continued series that in each remove differ very little one from the other. There are fishes that have wings and are not strangers to the airy region; and there are some birds that are inhabitants of the water, whose blood is as cold as fishes. . . . There are animals so near of kin both to birds and beasts that they are in the middle between both. Amphibious animals link the terrestrial and aquatic together; . . . not to mention what is confidently reported of mermaids or sea-men. There are some brutes that seem to have as much reason and knowledge as some that are called men; and the animal and vegetable kingdoms are so nearly joined, that if you will take the lowest of one and the highest of the other, there will scarce be perceived any great difference between them; and so on until we come to the lowest and the most unorganical parts of matter, we shall find everywhere that the several species are linked together, and differ but in almost insensible degrees. And when we consider the infinite power and wisdom of the Maker, we have reason to think, that it is suitable to the magnificent harmony of the universe, and the great design and infinite goodness of the architect, that the species of creatures should also, by gentle degrees, ascend upwards from us towards his infinite perfection, as we see they gradually descend from us downwards.

What is distinctive here is not Locke's insistence upon the continuity of nature but that his understanding of organic continuity was radically different from our own. The *scala naturae* was not a taxonomic device for expressing the continuity between animate and inanimate nature as a modern biochemist might understand it. Such chemical continuity was in fact denied, and the empirical evidence for the chemical nature of life would not begin to be available until 1828, with the first laboratory synthesis of an organic compound, urea.[15] The general acceptance of the chemical continuity of all nature, even among scientists, would not take place until the end of the nineteenth century. Locke was expressing something very different from a scientific conception of nature, and the extent of this difference is betrayed by his last line, "that the species of creatures should also, by gentle degrees, ascend upwards from us towards his infinite perfection, as we see they gradually descend from us downwards." The chain of being was not a taxonomic principle, but a metaphysical one; and it is less a *scala naturae* than a scale of perfection, extending down to base matter and up to the eternal incorporeal reality of the divine mind. Epistemologically, its roots are one with Plato's myth of the cave, in which the world of matter is but a dim and imperfect reflection of the eternal, unchanging idea. Matter, in such a metaphysic, is a reflection or substrate of eternal form.

There were critics of the idea, to be sure, Samuel Johnson among them, whom we can almost see kicking the pavements as he expostulates the following: "This scale of Being I have demonstrated to be raised by presumptuous Imagination, to rest on Nothing at the Bottom, to lean on Nothing at the Top, and to have Vacuities from Step to Step through which any order of Being may sink into Nilhity without any Inconvenience, so far as we can Judge, to the next Rank above or below it."[16] But Johnson's opinion was not the historically prevailing one, and it is certainly not the opinion to be found in William Smellie's textbook, *Philosophy of Natural History,* where one chapter is entitled "Of the progressive scale or chain of beings in the universe."[17] When we read Buffon's article on the ass, the intellectual context of the argument is that so aptly summarized in 1760 by the philosopher Robinet: "Le prototype est un principe intellectuel qui ne s'altère qu'en se realisant dans la matière."[18] (The prototype is an intellectual principle which in material form becomes debased.) Although this essentialist metaphysics provided the concept of a species to eighteenth-century biology, it also regarded the morphological variability of actual populations as a degradation of eternal form and the species themselves as ranked on a scale of perfection. Yet the static conception of nature implied by the essentialist theory of species was already being undermined in Buffon's generation by the growing appreciation of the magnitude of time.

THE DISCOVERY OF TIME

Only one year after Rousseau wrote the *Discourse on Inequality* for the contest at Dijon, Immanuel Kant, in a little-read book, *The Universal Natural History and Theory of the Heavens; or an Essay on the Constitution and Mechanical Origin of the Whole Universe Treated According to Newton's Principles,* argued that matter would form into solar systems and solar systems into clusters and clusters into larger clusters on a time scale hitherto unprecedented.[19] His argument was, to a modern scientist, an odd mix of both the metaphysical and astronomical, but there can be no doubt that there is a new awareness of the immensity of time. In his own words, "Millions and whole myriads of centuries will flow on, during which always new worlds and systems of worlds will be formed. . . ." This growing awareness of time and history was not confined to a single field, but its discoverers could range widely across the breadth of academic knowledge. If Kant produced a theory of the heavens later rediscovered by Herschel and Laplace, he could also give the most scientific account of the origin of human races.[20] If Laplace later ventured to trace the history of the universe, he could cite the paleontological research of Cuvier to make his physics more persuasive (Green 1959; Toulmin and Goodfield 1965).

It was James Hutton, Kant's close contemporary by age and inclination, who applied the levelers of time and change to the history of the earth. In *Theory of the Earth,* published in 1788, the year Buffon died, Hutton created the basis for Lyell's uniformitarian doctrine, which by Darwin's own admission was a direct intellectual precursor of the theory of organic evolution (Green, 1959; Toulmin and Goodfield 1965). In the humanities, too, the power of the past was making itself felt. The French and English colonization of India had made European scholarship aware of the ancient literary tradition of the Indian subcontinent, and there began the systematic program of translation and transcription which would give to the West a substantial body of Oriental literature by the end of the next century. In 1786, only two years before Hutton, Sir William Jones compared Sanskrit to Greek and Latin and drew his startling conclusion: "No philologer should examine all three, without believing they have sprung from some common source. . . ."[21]

Also during the eighteenth century, antiquarians undertook the excavation of Pompeii and Herculaneum, and scientists began voyages of exploration to collect both specimens of natural history and record ethnographic descriptions of native life.[22] In this endeavor, the romantic concept of the noble savage, much ridiculed by historians, had an important place in the development of anthropological decentration (Hallowell 1965). The concept of the noble savage questioned the entrenched view of savages as abject, miserable beings, necessarily inferior to Europeans. Moreover, it is nowhere as common in firsthand descriptions as some intellectual histories would lead one to believe; and in accounts of the Pacific region, it is a phenomenon of secondary sources.[23] Also, the idealistic notion has often been confounded with an ethical tradition of French letters, reaching from Montaigne to Diderot, which used the "noble" behavior of savages as a counterfoil to the behavior of Europeans.[24] The force of this argument derived precisely from the fact that savages were commonly considered to be inferior.

The ethnological decentration of modern anthropology derives from this humanistic critique, but its theoretical impact was minimal. Although the ethnological fieldwork method was first attempted by the students of Linnaeus (Blunt 1971, p. 186), the new facts of human variation and culture were commonly incorporated into a static conception of man and nature which had already been accommodated to the magnitude of time. In what Lovejoy calls the *temporalization of the great chain of being,* the hierarchical ranking of forms is given a chronological dimension, in which lower forms emerge first and higher forms last. It is a view of nature and history that is fundamentally static and typological in character, in which the emphasis is not at all on the process of change but on the enduring nature of the emergent forms.[25]

If the concept of successive emergence of forms develops within a metaphysical ranking of forms on a scale of perfection from lowest to highest, the result is not a modern theory of evolution but a temporal rendering of hierarchy in which nature and history are necessarily progressive. Nowhere is this clearer than in the work of Cuvier, the father of vertebrate paleontology, who gave us the actual bones of species hitherto unimagined and established without a doubt the reality of extinction. Cuvier (1813) refused to interpolate across the breaks in the fossil record but postulated waves of extinctions followed by the separate creation of distinctive new faunas by the hand of God. The temporalized chain of being retained immutability in the face of the magnitude of time.

Anthropological facts were also assimilated to temporalized hierarchy, and by the last decades of the eighteenth century, savages are once again on the bottom of the chain, intermediate between man and monkey, but with no more claim on a redeeming social equality or primitive contentment. The *scala naturae* did not disappear. Instead it acquired a cultural component and was laid on its side. No longer did it ascend upward toward infinite perfection but now moved forward toward a happier complex future while looking back at a simple, brutish past. What remains unchanged is the ethnocentric concept of the savage as a type of man. The savage is still inferior, but now this inferiority is secularized as fixation at an early stage of the historical succession. Given this interpretation, the application of the comparative method in anthropology became the specification of the characteristics of each historical stage.

This methodology, most elaborated in the late nineteenth century, was not the creation of Victorian anthropology but its inheritance. The examination of living savages to fathom the European past was commonplace by the beginning of the nineteenth century, and implicit in it were two assumptions: first, that the history of all cultural factors paralleled changes in technology and, second, that European society was therefore the most advanced sociocultural form. It was a method that made technology the measure of all things; and those who had dared, like Rousseau in the eighteenth century, to question the moral worth of technical progress and to raise the possibility that savages might be better off than Europeans became victims of ridicule that continues to this day. As early as 1826 the Eurocentric view of man had taken on a delusional character. Auguste Comte formulated stage-successive laws of history between bouts of clinical madness, and Manuel reports that "as his illness became aggravated, he had felt himself regress through various stages of metaphysics, monotheism, and polytheism, to fetishism, and then, in the process of recuperation, had watched himself mount again through the progressive changes of human consciousness, at once historical and individual, to positivism and health."[26]

Herbert Spencer, with less panache, had a similar realization in reading Lyell's refutation of Lamarck, and all of his subsequent writings were a working out of the view that all change is from the homogenous to the heterogenous based on differentiation of parts and from the lower to the higher (1852, 1857). Though Spencer had some appreciation of historical change, he had almost no interest in actual history itself but considered it pernicious to what he termed "a true theory of humanity."[27] His evolutionist scheme had been brought to bear on the facts rather than developed out of them. Lamarck's evolutionism is also a temporal rendering of hierarchy, as is clear from his own remarks; and though he wrote his major works in the Napoleonic era, his intellectual debt is to the generation who made the revolution.[28] In both Lamarck and Spencer, time is extraneous to the argument; and the temporal succession loses no logical continuity if collapsed to the synchronic plane.[29]

THE CHAIN BRANCHES

The comparative and evolutionary method that came to prominence in the anthropology of the late 1860s, with the publication of a number of synthetic books on savages and history by Bachofen (1867), Maine (1897), Morgan (1864), McLennan (1876), and Tylor (1889), was thus a working out of assumptions that owed nothing to biological evolution, in a modern sense of that concept, and it developed from a different tradition. Although all these books postdated Darwin's *On the Origin of Species,* their assumption that the characteristics of non-Western peoples (including such intellectual phenomena as language, descent systems, religion and art) represented merely an earlier stage in a progression toward European sociocultural forms is an assumption that long antedated the idea of the mutability of species and is logically independent of it.[30] Although Darwinian evolution, the historical change in the morphology of species under the operation of natural selection, was used in the late nineteenth century for the rationalization of avarice, a movement known as Social Darwinism, biological evolution had little effect on the anthropological image of man.[31] Darwin himself must share part of the blame for this, for he had as much difficulty in analytically separating biological and cultural evolution as any of his contemporaries. He not only took Wallace's advice that he use Herbert Spencer's phrase, "survival of the fittest," thereby forging a link between later editions of *On the Origin of Species* and the *Synthetic Philosophy;* but there is that strenuous subtitle to the *Origin* itself—"the preservation of favoured races in the struggle for life."[32] He may have noted in the margin of his copy of *Vestiges of Creation* to "never use the word 'higher' or 'lower,' " but he sometimes forgets his pledge in *The Descent of Man:* "The strong tendency of our nearest allies, the monkeys, in microcephalous idiots, and in the barbarous races of mankind, to imitate whatever they hear deserves notice, as bearing on the subject of imitation";[33] or further on: "Apes are much given to imitation, as are the lowest savages" (C. Darwin 1901, pp. 133, 198). Nonetheless, there are fewer such statements in Darwin than in the works of many who came later, and he was still too much the scientist to become a convert to Spencer's *Synthetic Philosophy*. He claimed that Spencer had no influence on his own work and deplored his "deductive manner of treating every subject" (Barlow 1958, pp. 108–109; Duncan 1911, pp. 87, 125; Freeman 1974).

As important as the theory of organic evolution may be for biological science, a theory that has its evidence in fields as diverse as embryology, biogeography, genetics, and anatomy, its importance to Western thought was immediately recognized as metaphysical (Carter 1957, pp. 64–67; Mayr 1972; Dewey 1910). Darwin's originality was to provide a mechanism that explained design in nature as the product of chance. Marx and Engels saw the affinity of a materialist, transformist biology to their own dialectical conception of history;[34] and when Ernst Haeckel popularized Darwinism on the Continent, he argued that its significance was the replacement of a teleological biology by a mechanical one. When Samuel Butler attacked Darwin in *Evolution: Old and New*, his objection was not to the mutability of species but to the mechanism of natural selection. This new evolution, by eliminating purpose in nature, eliminated even the deistic proof for the existence of God, the age-old argument from a design — that purpose in the creation implies the design of a maker (Ghiselin 1969). Even Asa Gray, professor of natural history at Harvard, Darwin's American bulldog, who defended evolutionary theory against its New World critics, tried to save the argument from design by the strategem that the variations on which selection acted were themselves purposive (Gray 1963; p. 127; Darwin 1862). Darwin attacked the argument from design directly in his monograph on orchids, which demonstrates that the organs of these plants were not originally formed for the functions they now perform. In Darwin's theory, even the most complex machines in nature, machines such as the human brain or the vertebrate eye, were evolved through the incremental selection of adaptive variations. Darwin's theory most directly challenged the Western image of man himself, which was still theistic in a way inanimate nature was not, and it was clearly incompatible with a scale of perfection.

The orthodox position, still widely accepted in Darwin's day and in our own, was that reiterated by Buffon in his essay on the nomenclature of apes:

> If our judgement were limited to figure alone, I acknowledge that
> the ape might be regarded as a variety of the human species. The
> Creator has not formed man's body on a model absolutely different

from that of the mere animal. He has comprehended the figure of man, as well as that of all other animals, under one general plan. But, at the same time that he has given him a material form similar to that of the ape, he has penetrated this animal body with a divine spirit. If he had conferred the same privilege, not on the ape, but on the meanest, and what appears to us to be the worst constructed animal, this species would soon have become the rival of man: it would have excelled all the other animals by thinking and speaking. Whatever resemblance, therefore, takes place between the Hottentot and the ape, the interval which separates them is immense; because the former is endowed with the faculties of thought and speech. Who will ever be able to ascertain how the organization of an idiot differs from that of another man? Yet the defect is certainly in the material organs, since the idiot is likewise endowed with a soul. Now, as between one man and another, where the whole structure is perfectly similar, a difference so small that it cannot be perceived is sufficient to prevent thought, we should not be surprised that it never appears in the ape, who is deprived of the necessary principles.[35]

This immortal soul gave man a unique place in nature, halfway between animal and spirit (Pope 1734, vol. 2, no. 2, pp. 3–10). Orthodox Christianity was fundamentally dualistic in its conception of the relationship of man to nature. The body of man was constructed in a way similar to the bodies of other animals, and man shared with them a number of animal passions and suffered the same range of animal infirmities; but his humanity consisted not in what he owed to nature but in what he owed to God. The human soul was immortal and participated in the divine even while encapsulated in matter, and creation of the human soul was a direct act of God. Direct divine intervention not only was a metaphysical supposition but intruded into the science of the day as the explanation of the origin of species. When Darwin put forth his hypothesis that species were formed by natural selection, as domestic varieties are created by selective breeding, he did not mention man in the book, save for acknowledgment of his role in selective breeding and a brief comment on the status of separate races (Darwin 1966). But Thomas Henry Huxley, in *Man's Place in Nature*, mustered all of the anatomical evidence for the similarity of man and ape that had been accumulating since the seventeenth century and drew the inevitable conclusion (1959, p. 125):

> But if Man be separated by no greater structural barrier from the brutes than they are from one another — then it seems to follow that if any process of physical causation can be discovered by which the genera and families of ordinary animals have been produced, that process of causation is amply sufficient to account for the origin of Man. In other words, if it could be shown that the Marmosets, for example, have arisen by gradual modification of the ordinary Platyrhini, or that both Marmosets and

Platyrhini are modified ramifications of a primitive stock — then, there would be no rational ground for doubting that man might have originated, *in the one case, by the gradual modification of a man-like ape; or, in the other case, as a ramification of the same primitive stock as those apes.*

The theory of evolution also ultimately destroyed the chain of being in biology, for what had formerly appeared as a hierarchy of morphological complexity was now interpreted as a number of distinct lineages seen on a single plane of time. The link between similar organisms was no longer a metaphysical link between species existing at the same time but a causal link that had once existed in the past but which was no longer present. In Darwinian biology, a branching tree replaced the chain of being, and anatomical homology was reinterpreted as the expression of an actual genealogical relationship and not simply as a static geometry of differences (Mayr 1963; Simpson 1953). Paradoxically, however, the unity of living matter and the continuity of species implied by Darwin's theory did not lead to a repudiation of dualistic theories of man and nature but to an absolute distinction between biological and cultural evolution. When the superstructure of Victorian thought was dismantled about the turn of the century, it did not give way to an evolutionary theory of man but to an evolutionary interpretation of the human body and to a historicist theory of human action. An integrated approach, then as now, was forestalled by far more immediate political concerns of the polemics and practices of racism.

THE NEW DUALISM

Darwin had an immediate impact on the anthropological study of race, if not on the theory of society, for biological evolution made it unlikely that the different human races had been separately created by God. The new evolutionism was on the side of the monogenists who had long argued for the biological unity of the human species.[36] Like most propositions in the study of man, the unity of the human species was politically controversial, with the advocates of slavery maintaining that Negroes had been separately created for their servile condition. This political division was reflected in the institutionalization of anthropology itself (Burrow 1966, pp. 234–235; Stocking 1971).

The Ethnological Society of London, founded in 1843, betrayed its humanistic roots in the fact that it was derived from the Aboriginal Protection Society, which had as its motto the then-controversial phrase, "of one blood"; and the Anthropological Society of London broke away from this parent organization in the 1860s, in part over differing interpretations of the racial question. The founder of the Anthropological Society, James Hunt, wrote a tract called *The Negro's Place in Nature* (1865), an obvious jibe at

Huxley, whose content can be deduced from its title; and meetings of the society's inner circle were brought into session by tapping a gavel fashioned to resemble a Negro's head (Stocking 1971, p. 380). Racism was long implicit in the treatment of black by white and had earlier been justified on religious grounds (Jordan 1969); but modern racism is secular racism, for a scientific age does not rationalize its prejudice by reference to the sons of Ham. The Enlightenment's interest in developing a natural history of man resulted in a biological classification of races, based on comparative anatomy, and led to a general recognition of the biological variety of the human species. Yet, the biological features of populations, such as the Oriental eye fold, and cultural features, such as eating with chopsticks, which we now call *racial* and *ethnic* differences respectively, were not explicitly distinguished until the twentieth century. Victorian scholars used the term *race* indiscriminately for both. In this too they were following the usage of Enlightenment anthropology, which had begun its ethnographic description with composite notes on both the physique and customs of the South Sea Islanders. In the eighteenth century, differences in human groups were most commonly explained in terms of a theory of climatic determinism, popularized by Montesquieu and the French Enlightenment, and reaction took the form of emphasizing the biological differences among populations, with no clear distinction among cultural achievements and innate abilities (Montesquieu 1900; Buffon 1797 or 1866). J. R. Forster (1778, pp. 227, 247), naturalist on Cook's second voyage, was eulogizing racial purity as early as 1778;[37] and David Hume, the year Rousseau wrote the *Discourse on Inequality,* argued that there was never a civilized nation of any complexion but white and that this superiority of the white race was due to a distinction in nature.[38]

When the racial typologies of anatomy were joined to the stage-successive theory of history, the technical backwardness of savages could be seen as the product of a more fundamental deficiency of intellect. Tylor, with his Quaker origins, may have accepted the "psychic unity of mankind" as axiomatic to anthropology, even though accepting historically conditioned differences between civilized and primitive thought, but other scholars pursued a stage-successive psychology to its logical conclusion, very explicitly set forth by W. J. McGee, who had read his Spencer, in *American Anthropologist* in 1901 (p. 238):

> The modern platform for the study of psychic homologies may be defined briefly in terms of a few generalizations, which seem to be consistent with the sum knowledge concerning the psychic attributes of both human and sub-human organisms, viz: (1) the mentality of animals is instinctive rather than ratiocinative, and for each species responds practically alike to like stimuli; (2) the savage mind is shaped largely by instinct, and responds nearly alike to like stimuli;

(3) all barbaric minds are measurably similar in their responses to environmental stimuli; (4) civilized minds rise well above instinct, and work in fairly similar ways under like stimuli; and (5) enlightened minds are essentially ratiocinative, largely independent of instinct, and less nearly alike in their responses to external stimuli than those of lower culture. The several generalizations are mutually and significantly harmonious; they combine to outline a course of development beginning in the animal realm with organisms adapted to environment through physiologic processes, and ending in that realm of enlightened humanity in which mind molds environment through nature-conquest; and they measure the gradual mergence of bestial instinct in the brightening intellect of progressive humanity.

By incorporating psychological factors into the stage-successive view of human history, late Victorian anthropology was able to construct a highly integrated theory, which could place savage man in the context of global changes in the methods of production, characterize the general features of his rudimentary culture, account for his backwardness in terms of insufficient rational control of instinctive tendencies, and show that some of his insufficiency was innate. Unfortunately for the theory, field anthropologists discovered that once the initial shock subsided of living with people who did not wear any clothes, savages were better characterized by what they shared with common humanity than by the ways in which they differed from idealized European intellectuals. Even worse, when the gross similarities in the methods of production were excluded, savages were found to differ as much among themselves as they did from Europeans.

The analytic separation of race and culture was taking place even as the integrated stage-successive theory was achieving its final form, and this distinction undercut the theoretical premises of racism and destroyed the heavily Eurocentric bias of Victorian anthropology. However, it also led to the postulation of two distinct levels of human functioning, a biological and a cultural level. This theoretically justified the implicit bifurcation of anthropology into physical anthropology, which dealt with man's body, and cultural anthropology which treated, not so much his soul, as that part of it which was acquired by man as a member of society. The distinction between race and culture was assimilated to the prevailing stratified conception of man, and culture now became an emergent property of organic evolution. Kroeber (1917) spoke of culture as "superorganic," and Goldenweiser (1935, p. 3) begins his text with the statement that man is "immeasurably removed from all his precursors," "a super-animal." The prefix *super*, especially in American English, is an implicit ideological assertion about the superiority of nurture over nature. It implies that there is a distinct level of phenomena, a cultural level, that is not integrated with human biology but is over and above it. Culture becomes a dividing line between man and animal, and an old dualism perpetuates itself in a new form.

Even Malinowski, who was no culturologist, and who wrestled with the integrative problem of biology and culture, is most revealing in the biological human nature he argues for: impulses for oxygen, hunger, thirst, sexual coupling, fatigue, restlessness, sleep, urination, elimination, fright, and pain. It is a biology that would characterize any vertebrate, and there is nothing distinctively human about it. Humanity, by implication, is a strictly cultural phenomenon (Malinowski, 1960). One consequence of such a conceptualization of culture is that human evolution is totally incomprehensible without postulating a functional discontinuity in the course of phylogeny as absolute as the faunal successions of Cuvier. The fossil record, however, shows an increase in cranial capacity through the Pleistocene.[39] Arthur Keith, remaining faithful to the logic of his assumptions, suggested that there was a "cerebral Rubicon" of cranial capacity, above which a human level of function emerged (1948, pp. 205–206).

In British anthropology,[40] the emphasis was more on society than culture, but it accepted an equally dualistic conception of the relationship between human nature and human society. This fission began as a methodological heuristic to describe social phenomena in purely social terms, without imputing motivations to the people involved; and, as a method, it made good sense in the context of Victorian social science, with its lists of instincts and its easy attributions of racial mentalities. W. H. R. Rivers, one of the first to argue for sociological autonomy, was not hostile to psychology; but he viewed the relationship between psychology and culture as hierarchical (1924). He argued that psychology was related to social science as physical science was related to geological science, as providing a substrate of laws on which historical process could act. That the segregation of disciplines was much greater than this analogy implies is best conveyed by Rivers himself, who could write a book called *Instinct and the Unconscious: a Contribution to a Biological Theory of the Psycho-neuroses* without once considering the implications of these phenomena for his autonomous sociology.

The recognition that psychology and anthropology have different methodologies and study different aspects of human activity does not require us to believe that man himself is divided into psychological and cultural components or that his culture and his mind can lead independent existences. This later interpretation, if not often explicitly stated, is certainly latent in the anthropology that came to prominence in the twentieth century. It is a view that is encouraged by Comte's conception of the hierarchy of the sciences, and the debt of social anthropology to the writings of Comte has been acknowledged by Radcliffe-Brown, another key figure in the attack on the historicism of Victorian anthropology.

Radcliffe-Brown, in the introduction to *Structure and Function in Primitive Society,* published in 1952, defines anthropology as the study of the processes of social life; but he goes on to accept a distinction originally made by Comte by analogy with the physics of the early nineteenth century: that there are two sets of sociological problems, *social statics,* which deal with the conditions for the existence of a social form, and *social dynamics,* which deal with the conditions for its change.[41] There are universities establishing departments of social change today, but historically such conceptions derive from a world view in which society is conceived of as an enduring form and history as a force that replaces one stage by another. British structural anthropology, which viewed itself as a radical departure from the evolutionism of the Victorians, was in many ways only its obverse. Both methodologies accepted the reality of societies as enduring forms, and both methodologies implicitly accepted a pre-Darwinian world in which structure and change are different types of phenomena. How else are we to interpret the quotation from De Saussure which prefaces Fortes's *Kinship and the Social Order?* "The opposition between the two viewpoints, the synchronic and the diachronic, is absolute and allows of no compromise" (Fortes 1969, epigram).

Clifford Geertz has attempted to put these developments of twentieth century anthropology into broader perspective, and he has pointed out that insofar as anthropologists searched for the common denominators of culture, they adhered to a "stratified" view of the relationship between biology, psychology, and culture that failed to recognize that the ethnographic method heralded a significant departure from the uniform human nature garbed in the cloak of custom (1975, p. 51). What has still not been appreciated, even by such thoughtful commentators as Clifford Geertz, is that the discovery of culture did not abolish the typological conception of human nature but perpetuated it. To say that human nature remains incomplete, if it does not interact with cultural information, may effectively abolish the concept of man in the state of nature — a concept which probably few educated people, least of all its creators, actually believed in any simple sense — but its ultimate effect is to deny that human differences in behavior are ever biologically caused. To argue that all behavioral variation among human populations is culturally induced is to assert by implication that human nature is everywhere identical. Yet such an assertion is contradicted by the entire thrust of twentieth century genetics, which has been to show that biological systems are inherently variable.

A witness to the triumph of autonomous cultural and social anthropology early in this century, Malinowski's professor at the London School of Economics, Edward Westermarck, clearly foresaw the danger of a social science based on a stratified conception of man (1921, p. 9, Introduction):

> For the present, then, we should on this principle, carefully refrain from assuming, for example, that courtship and marriage have anything to do with the sexual instinct, that the retaliation of adultery springs from jealousy and revenge, that the secrecy observed in the

performance of the sexual function is connected with sexual modesty. We should refrain from trying to find any motives for the practice of polygyny, the prohibition of incest, the various marriage rites, and so forth. We should only correlate these phenomena with other social phenomena or refer them to social antecedents. Dr. Rivers says, in fact, that it would be possible "to write volumes on that group of social processes which we sum up under the term 'marriage,' without the use of a single psychological term referring to instincts, emotions, sentiments, ideas or beliefs," and that such a treatment of marriage would nevertheless be "capable of producing valuable contributions to our knowledge." After those volumes had been written we might perhaps be allowed to consider that people not only marry but fall in love, and that the marriage customs are not merely muscular movements standing in relation to other muscular movements, but that they are actuated by intentions and motives. I should not be surprised, however, if social psychology, when at last permitted to speak, should raise violent objections to many of the classifications and conclusions made by "pure" sociology. For it seems to me that "pure" sociology is liable to commit the most fatal mistakes by detaching social phenomena from their motive powers and treating them as mechanical processes, just as if men as members of society were a sort of automata.

Social psychology did speak, but not with one voice, and it was not the ally that Westermarck had hoped for. To the contrary, it was implementing a program of methodological reform similar to that being pursued in anthropology by reacting against instinct as an explanatory concept in human behavior. William McDougall, reader in mental philosophy at Oxford, whose *Social Psychology* was often reprinted through the first two decades of this century, unbeknownst to himself, was fighting a rearguard action. Complaining that psychology was denied its rightful place as the foundation discipline of the social sciences, especially by Émile Durkheim, he foresaw that "this regrettable state of affairs is about to pass away" to be replaced by a universal recognition of the importance of a science of mind (McDougall 1920, pp. 1–2).

THE LADDER OF REASON

Psychology did become influential again in the social sciences but hardly in the way that McDougall has envisioned. Instead, John B. Watson brought psychology itself into conformity with the assumptions of Durkheimian sociology by the movement known as behaviorism. Although Watson had undertaken one of the first ethological field studies — of sea birds in the Caribbean — and argued for a wide range of instincts in both animals and men in his book *Behavior,* published in 1914, by the early twenties he stressed an extreme environmentalism that not only went beyond the data but contradicted what he had summarized in his own earlier works (1914, 1970). Watson argued for a methodological purge of mentalism in psychology that repudiated thought, consciousness, and instinct as unscientific concepts and replaced them with the study of overt behavior. Watson himself accepted the fact of unlearned responses in man, but his theoretical position postulated an opposition between learned and unlearned responses in which the learned emerged later during development and became the predominant form of activity. Most revealing is his chart of development in which the two types of behavior diverge in infancy and run into adulthood as separate lines (1970, p. 138). These two developmental streams — one later, malleable, dominant and better, the other earlier, fixed, subordinate, and not so good — reveal behaviorism to be a most reputable member of the Western intellectual tradition. The behaviorist program, as it developed after Watson, particularly in the United States, led to a severe contraction of the topics and species studies by psychologists, until it reached that unenviable state summed up in Frank Beach's wry definition of comparative psychology as the study of learning in the white rat and college sophomore.[42]

Instinct theory had been banished to the periphery, but it did not go away. It persisted in the studies of European ethologists, in psychoanalysis, in theories of emotion, and covertly, in learning theory itself, as the motivational drive and the unconditioned reflex. Even here, in those fields most committed to instinct theory as an explanatory concept, fields such as psychoanalysis, the moral devaluation of instinct was often as strong as in those schools that rejected it entirely, as in the pessimistic prognostication of the human future, laid at the door of instinct by Sigmund Freud in *Civilization and its Discontents* (1962).[43] Professional anthropology made a brief flirtation with psychoanalysis in the early decades of this century, but its general position has been repudiation of instinct as an explanatory factor in human behavior. In maintaining this thesis, it unwittingly accepted the major assumption that underlies both behaviorism and psychoanalysis: that learning and instinct are competing explanations of behavioral phenomena. By aligning itself with one side of this controversy, anthropology ensured that instinct theory would continue to dominate social theory, not as a positive, but as a negative. Westermarck and McDougall both failed to recognize that the autonomy of social facts, advocated by their nemesis, Émile Durkheim, was not the repudiation of psychological variables in social science but the substitution of a negative theory of instinct for the positive theory of their own. Émile Durkheim had a psychological theory, but it was a theory so crudely drawn that it successfully slipped past as no theory at all.

In the context of the late nineteenth century, which placed what seems to us undue emphasis on the possibility of being captain of one's soul and master of one's fate, especially in the face of the vast social forces unleashed by the rise of industry, it was reasonable for Durkheim to stress the limits of individual autonomy. Confronted with the elaborate institutions of the bureaucratic state, only then taking their modern form, it was reasonable to repudiate the utilitarian conception of society as an aggregate of individuals and to argue for a concept of social institution that was distinct from the individuals who participated in it. Yet in pursuing these reasonable ends, Durkheim did not repudiate the image of man held by the Victorians but simply perpetuated an inversion more congenial to his analytic needs. He accepted without question the moral devaluation of instinct and the supposed asocial nature of man but emphasized instead the residual character of the biological components of behavior in man and the power of socialization to curtail their undesirable tendencies. In Durkheim's sociology, the distinction between his major categories, mechanical solidarity and organic solidarity, is ultimately a difference in the mechanisms of social control (1964b); and in both cases it is the asocial nature of man which must be coerced and channeled. In *Rules of Sociological Method,* he actually argues that the term *social fact* be made coextensive with coercion by restricting it to only those social phenomena that the individual is powerless to affect (1964a); and in *Moral Education* he developed a theory of ethics in which the questions of right action that so long vexed philosophers are subsumed under a system of social control (1961). "Morality," he writes, "is a comprehensive system of prohibitions" (1961, p. 42). But further along the same page, it can be seen what lurks behind such a schoolmaster theory of ethics:

> The totality of moral regulations really forms about each person an imaginary wall, at the foot of which a multitude of human passions simply die without being able to go further. For the same reason — that they are contained — it becomes possible to satisfy them. But if at any point this barrier weakens, human forces — until now restrained — pour tumultuously through the open breach; once loosed, they find no limits where they can or must stop.

For this pessimistic view of human nature, Durkheim is quite explicit in stating his debt to the soul. He writes that every age has conceived of man as "formed of two radically heterogeneous beings: the body and the soul," and he concludes that "a belief that is as universal and permanent as this cannot be purely illusory" (1914, p. 328). He goes on to say that "psychological analysis has, in fact, confirmed the existence of this duality: it finds it at the very heart of our inner life": the contrast between "sensations and sensory tendencies" on one hand and "conceptual thought and moral activity" on the other. He gives this opposition a moral dimension: "Not only are these two states of consciousness different in their origins and their properties, but there is a true antagonism between them. They mutually contradict and deny each other. We cannot pursue moral ends without causing a split within ourselves, without offending the instincts and the penchants that are most deeply rooted in our bodies." This is no neutral sociology, no argued refutation of psychological constructs in social science, but a rendering in sociological terms of a profound anxiety about human feelings. It indeed owes its origins to the soul, and not simply to the facultative soul of rationalist philosophy but to the concupiscent human nature of the Judeo-Christian tradition. Durkheimian sociology is first and foremost a psychological theory of human nature and a psychological theory of instinct, and it is a theory strongly at variance with both a naturalistic approach to human society and a modern conception of instinct.

Underlying Durkheim's distrust of the passions and historically related to it is the equally enduring metaphysical opposition between instinct and reason already encountered in the anthropology of W. J. McGee, which makes instinct part of the animal nature of man and reason the human essence. However, as Johann Gottfried Von Herder, the younger cofounder of romanticism, had cautioned, in his *Essay on the Origin of Language*, submitted to the Academy of Berlin in 1770, there is an unfortunate tendency to "think of man's reason as a new and totally detached power that was put into his soul and given to him before all animals as a special additional gift and which, like the fourth step of a ladder with three steps below, must be considered by itself" (Herder 1966, p. 110).[44] In rejecting the necessary opposition between instinct and morality implicit in the Durkheimian theory of social control, modern ethology also rejects a necessary opposition between instinct and learning. Although ethology began with the study of behavior that was purely innate, in the sense that it appeared fully formed in animals that had been deprived of an opportunity to practice or to learn from others—behavior that has been empirically shown to exist—other behaviors have also been demonstrated which are not innate in this strict sense. However, the existence of behaviors that are not *strictly* innate does not require us to infer that they must be the opposite of innate. Learned behavior, in its role as the psychological underpinning of the socialization theory of modern social thought, has been interpreted as just such an opposite. Yet between the strictly innate and the strictly learned stands a whole spectrum of empirical possibilities, some of which have been demonstrated by ethological investigation.

Konrad Lorenz, in one of the first considerations of the problem of the interaction of innate and environmental information from the perspective of ethology, had this to say in the 1920s, when behaviorism was being devised (1957, p. 137):

> Instinctively innate and individually acquired links often succeed each other directly in the functionally uniform action chains of higher animals. I have termed this phenomenon "instinct-training interlocking," and emphasized that similar interlockings also occur between instinctive and insightful behavior. Here, where we are

dealing with the effects of experience, we must first discuss the interlocking of instinct and training. Its essence lies in the fact that a conditioned action is inserted in an innate chain of acts at a certain, also innately determined point. This action must be acquired by each individual in the course of his ontogenetic development. In such a case the innate action chain has a *gap*, into which, instead of an instinctive act, a "faculty to acquire" is inserted.

The concept of an interlocking or intercalating of instinctive and learned behavior, as developed by Lorenz, was a first step in overcoming the opposition between innate and learned behavior that characterized most American psychology, but it still presumed two distinct categories, which were parceled out sequentially in a common act. More recent research indicates that instinct and learning can actually combine.

The possibilities are revealed even among closely related species of birds (Kroodsma 1977; Nottebohm 1970, 1972; Marler and Peters 1977; Marler and Mundinger 1971). The calls of most birds are strictly instinctive, in that the normal call repertory develops in a bird that is deafened soon after hatching. This is true of the call repertories of both chickens and ring doves. However, some birds have an elaborated call known as the *song* which has a more complex ontogeny. The song sparrow (*Melospiza melodia*), though reared in isolation, will develop some of the normal components of species-specific song if allowed to hear itself sing; but if it is deafened its song develops even more abnormally. However, other species require exposure to a model song. If a male white-crowned sparrow *(Zonotrichia leucophrys)* is raised in isolation from other male birds, it will fail to develop a normal song and will produce an acoustically impoverished song as an adult. Traditional socialization theory would predict this result, but some other aspects of the phenomenon cannot be explained by socialization theory at all. For example, the song learning is bounded by a critical period, and the exposure to song must take place when the bird is primarily between ten and fifty days of age. Once this period is past, the bird is refractory to further learning and retains its song, normal or abnormal, throughout life. An even more significant fact is that the bird is selective about what constitutes a model. If the young male is exposed to white-crowned songs during the critical period, it will develop a normal adult song, but if it is exposed to the songs of alien species, it does not learn the alien songs but develops abnormal songs of the white-crowned sparrow type. In a related phenomenon, the isolated chaffinch (*Fringilla melodia*) will imitate chaffinch songs that have been experimentally altered, and it will imitate recordings of the tree pipit (*Anthus trivialis*), which has a similar song to the chaffinch; but it will not imitate more divergent songs. In contrast to these species, the zebra finch (*Taeniopigia guttata*), if reared by a foster parent of another species, will develop the song of its foster parent in some circumstances.

The adopted male zebra finch will develop an abnormal song if reared by a female Bengalese finch (*Lonchura striata*), with no exposure to male songs at all; but it will develop a Bengalese male song if exposed to a male Bengalese finch. It is of interest too that the adopted male zebra finch will learn the alien song of its alien foster father even if exposed to male zebra finch songs.

Although birds are phylogenetically remote from primates, they are the only other group of vertebrates, with the possible exception of whales and dolphins, known to have developed traditionally transmitted vocal communication. Like in man, too, their vocal productions are controlled by one side of the brain. As such, vocal learning in birds cannot be dismissed as too remote from man to permit comparison. The complex interactions between instinct and experience that have been demonstrated in song birds are an important contribution to our understanding of behavior generally. They convincingly show that the absence of strictly instinctive behavior can by no means serve as proof for the irrelevance of innate factors in behavioral acquisition. Although the innate and environmental components are conceptually distinct, in that they may be teased apart by the appropriate experiments, they function as an integrated unit in the animal's behavior. The mature male chaffinch does not parcel out his mating song into innate and acquired productions but produces a song that is the product of his developmental history. Bird song acquisition is most directly relevant to theories of vocal-auditory communication in man, and here the importance of innate components has long been recognized.

COUNTERATTACK FROM THE EAST[45]

When Noam Chomsky (1959, 1968) first attacked the behaviorist assumptions of social science in the late 1950s, he emphasized the inadequacy of a simple learning theory model to account for the complex constraints that characterize linguistic data. In making this argument, he revitalized a tradition of organic continuity in language studies that is as old as linguistics itself. The development of linguistics in the nineteenth century began as part of the reaction to the denigration of feeling by the rationalist philosophy of the French Enlightenment. This antirationalist movement, associated historically with Rousseau and Herder, questioned the moral claim of the emerging technological world and gave voice to a conception of the organic connectedness between man and nature and man and society. The forcefulness of this critique, which has informed conceptions as various as Karl Marx's attack on the alienation of labor and the modern ecologist's indictment of industrial pollution, is belied by the insipidity of its traditional label, the *romantic movement*. This term, which is most correctly applied to the decline of classical canons of aesthetics in the late eighteenth century, nonetheless encapsulates a shift in polarity in the moral evaluation of the primitive, which had immediate implications for the nascent science of man. The romantic movement argued for the value and aesthetic primacy of those folk cultures and ancient customs considered to be outmoded by the theoreticians of technical progress. It enkindled a new enthusiasm for the Middle Ages, for the simple life, and for the cultures of past times. Moreover, it was associated with a rising German nationalism that elevated the Germanic languages to the same status as Latin and Greek and repudiated the Gallicized culture of eighteenth century Germany typified by the court of Frederick the Great. The systematic and descriptive study of languages, which we now call *linguistics,* was a direct descendant of the Teutonic attack on the French apotheosis of reason. However intuitive its goals, linguistic science began with the systematic application of the methods of comparative anatomy to human language.

As already noted, the concept of an Indo-European family of languages, distributed from the British Isles to eastern India, from Iceland to the Persian Gulf, was first developed in the eighteenth century, when Sir William Jones, a friend of Samuel Johnson, was sent as a judge to Calcutta.[46] Jones, then in his mid-thirties, was already a man of enormous linguistic erudition. A Fellow of the Royal Society at twenty-six, he had mastered Latin, Greek, French, and Italian before leaving Harrow, acquired some facility in Spanish, Portuguese, German and Arabic at Oxford, and published a grammar of Persian not long after. A month after taking up his legal duties in Bengal, he founded the Asiatick Society, in 1784, and a year and a half later began the study of Sanskrit. His wide knowledge of languages allowed him to perceive the resemblances between Sanskrit, European languages, and Persian, but most crucially, he drew a Darwinian conclusion: he did not say that Sanskrit was ancestral to European languages, the conclusion later drawn by Friedrich Von Schlegel, but that all of these languages were derived from a common source. Jones, who had translated Persian odes into French and written a treatise on oriental poetry in Latin, did not dismiss the sacred language of the Hindus as a primitive survival nor its bearers as lesser breeds without the law. His vision of the past was not the stage-successive history of primitive and advanced but the branching tree, which he first applied in linguistics nearly eighty years before *On the Origin of Species.*

His method was comparative:

> The *Sanskrit* language, whatever be its antiquity, is of a wonderful structure; more perfect than the *Greek,* more copious than the *Latin,* and more exquisitely refined than either, yet bearing to both of them a stronger affinity, both in the roots of verbs and in the forms of grammar, than could possibly have been produced by accident.[47]

At the time that Jones wrote there had already been several centuries of European linguistic comparisons, and it was this tradition which developed standards for dating and authenticating texts. By the early seventeenth century, for example, it had already been shown that the writings of the supposed pre-Greek magician, Hermes Trismegistus, which had greatly influenced Copernicus and Bruno, were in fact of Roman date and not ancestral to classical philosophy.[48] Its more arcane forms, however, led to rampant homologizing on the basis of superficial resemblances of words, homologies which could always be found since European languages borrowed and reborrowed words down through the centuries, diverse languages share features of phonetic symbolism, and the sounds of one language may converge toward those of another. In contrast, it was the systematic comparisons that caught Jones's eye, those that could not possibly have been produced by accident. The subsequent linguistics of the nineteenth century succeeded where the early philologists failed because they developed methods of systematic comparison that achieved the distinction known to biologists as the difference between homology and analogy: homology being a similarity based upon descent from a common ancestor, analogy a convergence toward a similar form among unrelated structures.

Sir William Jones died without ever returning to Europe, but his abiding interest in the Sanskrit language was shared by an ensign in the army of the East India Company, Alexander Hamilton, who became the historical link between the Asiatick Society of Bengal and the subsequent efflorescence of nineteenth century linguistic scholarship. Sometime between May 1802 and April 1803, Hamilton traveled to Paris; and since he was reputed to have been the only person on the Continent who knew Sanskrit, he was asked to prepare a catalogue of the Indian manuscripts held by what is now the Bibliothèque Nationale. Before returning to a quiet life as professor of Oriental languages at the East India Company's training school in England, he gave instruction to Chezy, who later became the first professor of Sanskrit on the European mainland, and lived in a house with Friedrich Von Schlegel, romanticist philosopher, who later published *The Language and Wisdom of India* in German in 1808. This book disseminated the insight and enthusiasm of William Jones to German-speaking Europe and advocated a study of language based on the methodology of comparative anatomy. The book was credited by Franz Bopp, one of the most erudite Indo-Europeanists that the century would produce, with inspiring Bopp, as a young man, to travel to Paris to undertake the study of Sanskrit with Chezy.

It was Bopp who in 1816 published the comparative grammar of languages envisioned by Schlegel and who demonstrated through detailed explication of the correspondences in Greek, German, Latin, Persian, and Sanskrit the validity of William Jones's original surmise. Bopp agreed with Jones, and contradicted Schlegel, that all of these languages had sprung from some common source. At the same time that Bopp was pursuing his Sanskrit studies in Paris, another linguistic polymath, the Dane, Rasmus

Rask, was in Iceland perfecting his knowledge of Icelandic. Rask had already written a book on the subject, *Introduction to the Icelandic or Old Norse Speech,* which he had finished when he was twenty-two, but the new venture was in response to a contest organized by the Danish Scientific Society on the origins and connections of "the old Scandinavian speech." In preparation, Rask had already learned Swedish, Finnish, and Lapp. His monograph, which was not published until 1818, compared Icelandic primarily to Latin and Greek with observations on Baltic, Slavic, and Germanic languages, and showed that Icelandic and the classical languages must be distinguished from such languages as Finnish, which are now placed by contemporary linguists in the family Finno–Ugric. Rask was also the first person to point out the systematic differences in sounds among Latin, Greek, and Germanic languages, later known as Grimm's Law.

At the time Rask was performing his Icelandic researches, Jacob Grimm had already produced a comparative study of the Germanic languages and their historical changes, as deduced from surviving texts. In the second edition of his compilation, in 1822, he incorporated a modified version of Rask's observations on systematic sound change in European languages, expanded to include Sanskrit. He pointed out for example, where Greek, Latin, and Sanskrit had *p,* Germanic had *f.* Thus, in the word for "foot," the cognate forms are *fotus* (Gothic), *podos* (Greek), *padas* (Sanskrit), and *pedis* (Latin). The establishment of homologous words and the characterization of historical sound change through the systematic comparisons of different languages became a minor scholarly industry in subsequent decades of the nineteenth century. It eventually came to be realized that the codification of such sound shifts was in effect a hypothesis about the relationships among languages, and by assuming that sound changes were always systematic, the method could be generalized to include all languages whether or not there were historical texts in existence. The aim of the comparative method, as developed by the school of German linguists known as the "young grammarians," became the reconstruction of a hypothetical source language, based upon the establishment of cognate forms in different languages that could be related to each other on the basis of systematic sound shifts over time. This development altered the concept of cognate words as understood by the eighteenth century. Joseph Banks, for example, the naturalist on Cook's first voyage, had collected word lists in Polynesia and compared them to other word lists obtained on the island of Madagascar in the Indian Ocean, half a world away.[49] He had concluded that the languages of Madagascar are related to the languages of Polynesia (which they are). However, no linguist a century later would have been convinced by such a superficial correspondence, for if correspondences are too exact, they are more likely to reflect borrowing than descent, since words derived from a common source language will not be the same but rather *systematically different*. The genius of nineteenth century linguistics was to demonstrate the empirical force of an almost contraintuitive insight: that two words, although they neither mean the same thing nor sound alike, can nonetheless be the same word. In arguing that sound change is systematic over time, comparative linguistics discovered the closest thing in social science to that most improbable of rationalist conceits, a law of history.

As the Sanskrit scholar William Dwight Whitney pointed out later in the nineteenth century, the creation of historical linguistics was primarily a German undertaking; and it is clear that systematic sound change is a somewhat paradoxical result (Whitney 1883, pp. 318–319; Silverstein 1971). Unlike Comte and the positivists, the founders of historical linguistics did not set out to exemplify the laws of history but were closely connected to its ideological opposite, the romantic movement. Friedrich Schlegel, who had first used the term *comparative grammar*, was one of the original formulators of the romanticist aesthetic, and he later repudiated the philosophical and political radicalism of his youth and converted to Roman Catholicism. His elder brother, August Wilhelm, though most famous for his role in German letters and his enthusiasm for the restoration of medieval literary themes, also set up the first Sanskrit printing press on the Continent and produced editions of the *Bhagavad Gita* and the *Ramayana,* with a translation of the latter into Latin. Jacob and Wilhelm Grimm, in conjunction with their linguistic research, began a massive compilation of Germanic folk traditions, of which Grimm's *Fairy Tales* are the best known, and published a nationalistic theory of Teutonic mythology, which exemplified the innate wisdom of the German *Volk*. Moreover, one goal of Jacob Grimm's linguistic researches was to show that the German language had once been as morphologically complex as Greek and Latin (Hjelmslev 1950–1951, p. 192).[50] Rasmus Rask's interest in Icelandic grew out of his boyhood ambition to found an ideal commonwealth in New Zealand based upon the Norse sagas. He was greatly influenced by the biologist Linnaeus: not only by the Linnaeus of the botany books, who codified nature, but by that romantic Linnaeus who had himself journeyed to Lapland as a youth and whose portrait shows him dressed in a Laplander's clothes.

Franz Bopp, not overtly romantic, nonetheless incorporates the reversed polarity of the primitive into his theory of language. Like August Schlegel, he believed that the process of linguistic change is not a neutral evolutionary transformation, as later linguists would think of it, but degeneration from primitive perfection. The languages known to us today "have lost more or less of what belonged to that perfect arrangement, in virtue of which the separate members were in accurate proportion to each other, and all derivative formations were still connected, by a visible and unimpaired bond, with that from which they originated" (Bopp 1974). In fact, Indo-European languages *have* simplified.

Bopp counted among his most intimate friends the Prussian ambassador to France, Wilhelm von Humboldt, who shared with him an organic view of language. Humboldt, whom Chomsky (1968, p. 15) has credited with an anticipation of generative grammar, argued that language was an expression of the genius of a people. He advocated a modern concept of a nation as a political entity based upon a unitary culture, of which language was the quintessential form. Moreover, in his treatise on language, published posthumously in 1836, Humboldt expresses a nonprogressivist concept of language that contrasts sharply with the guiding assumptions that came to prominence in anthropology a half century later (Humboldt 1971). Instead of regarding Chinese and Sanskrit as stages of linguistic evolution, he argued instead for independent paths of language development, sustained by the innate creative power of the human mind in culturally diverse form.

The close link between language and theories of organic continuity and between technology and theories of progress is not accidental. The languages of savages are not simple, as anyone who has tried to learn them has discovered. Even though all languages are not "equally complex," as shown by studies of acquisition, the discrepancies among levels of technological complexity, which are historical facts that must be reckoned with, are not paralleled by an equal range of discrepancies in structural complexity or semantic power among the world's languages. This fact is noted by W. D. Whitney, in a linguistic text contemporaneous with the ethnology bloom of the 1870s. Whitney attempts to explain this discrepancy between language and technology in technological terms (1883), p. 282):

> The essential difference, which separates man's means of communication in kind as well as degree from that of the other animals, is that, while the latter is instinctive, the former is, in all its parts, arbitrary and conventional.

The arbitrariness of language, as Whitney correctly saw, is supported by the general absence of any consistent relationship between speech sounds and meaning across languages, excepting phonetic symbolism, and the conventionality of language is shown by the fact that children will learn whatever language they are raised with. "As we approach man," he writes, recapitulating eighteenth century psychology, "the general capacities increase, but the specific instincts, the already formed and as it were educated capacities decrease. . . . Except suckling, he [man] can hardly be said to be born with an instinct" (1883, p. 289). This approach to language had a lineal descendant in another American linguist, Leonard Bloomfield, who in the 1930s introduced into the field a behaviorist conception of language acquisition and semantic theory (1962). This environmentalist emphasis in linguistics, which paralleled similar developments in the sister disciplines of psychology and anthropology already discussed, was nonetheless never reconciled with the anomalous facts of historical linguistics: the systematic nature of sound changes over time. In view of the close institutional relationship between anthropology and linguistics in American universities, reflected in the language courses included in the anthropology syllabus, it is odd to discover that the post-Victorian reaction against the concept of cultural homologies culled from the literature on savages did not lead to a similar repudiation of nineteenth century claims on the historical connectedness of modern languages. Why should the comparative method of anatomical homology, found wanting as a guide to history when applied to all other forms of social facts, reveal genealogical relationships among languages? How is it that historical linguistics can be possible at all? The implication of this question is that not all cultural phenomena are mediated by the same neural mechanisms but interact in various degrees with innate constraints. Psycholinguistic research in fact suggests that phonemic perception is mediated by innate feature-detectors or sequence detectors in the human auditory system which are perhaps species-specific for man (Denes and Pinson 1963; Liberman 1970; Lieberman 1975, chaps. 1–3; Stevens 1975)

As behaviorist social science was becoming institutionalized in the 1940s, the Haskins Laboratories in New York were beginning a research program that led to the systematic study of phoneme perception in speech. The success of this inquiry was made possible by the invention of a machine called the *pattern-playback*. When a sound spectrogram is inserted into this machine, it reproduces the sound represented by the graph. The advantage of this technology is that it allows portions of natural sounds to be progressively pared away or progressively added to in order to discover which portion of a sound carries linguistically significant information. This research has shown that there are some highly specialized aspects of human phoneme perception.

Phonemes constitute a class of speech sounds that are perceived as equivalent by a speaker of a language, like the *t* of *pat* and *top*. Acoustically, there are distinguishable differences between these sounds, but English speakers perceive both of these *t*'s as *t* and do not ordinarily distinguish between them. The repertory of phonemes in a language is used to characterize the differences among higher-level units in the language in a systematic way: *tap, pat, top, pot, pop, tat*, and so on are all potentially meaningful units in English built out of the concatenation of a small set of basic phonemes, in English about forty. However, psycholinguistic research has shown that there is no corresponding class of acoustic events that maps onto the phoneme repertory in a one-to-one manner. The human voice does not produce a sequence of acoustic cues that correspond to a sequence of phonemes but rather produces complex acoustic shapes that are broken down into phonemic sequences by the human ear. In other words, if a magnetic tape recording is made of the utterance "tap your fingers," it would not be possible to cut the tape into sections that correspond to *t*, to *a*, to *p*, because these phonemes are encoded into a complex wave form that requires the listener to reconstruct them.

Obviously, there must be acoustic features of the wave form that allow the listener to reconstruct one phoneme and not another, but the cues themselves are context-sensitive. The phoneme /d/, for example, is not represented by an invariant, acoustic shape or even by a family of such shapes of variable frequency but by a class of acoustic events that are physically very different from each other. Moreover, the phonemes are not separately encoded in the speech wave form but are coencoded such that the "cue" consists of a relationship among multiple features, as for example the cue for /d/ will vary as the following vowel varies. In figure 1.2, when the vowel is [i], as in [di], the /d/ phoneme is conveyed by the sudden rise in frequency shown by the "tail" or formant transition as circled. When the following vowel is [u], as in [du] in figure 1.2, the /d/ phoneme is conveyed by a sudden *fall* in formant frequency. However, it would be incorrect to conclude that these formant transitions of rising or falling frequency were the cues for the phoneme /d/ because exactly the same transitions can signal other phonemes in other acoustic contexts. In fact, it would not even be correct to call these formant transitions "speech" at all because if they are played by themselves on the pattern-playback, without the vocalic context, they do not sound like consonants in speech but like the warbles of bird song. Phonemes are a psychological fact, confirmed by the way in which speakers of a language will hear physically disparate signals as the "same thing" and combine these classes into semantic units in systematic ways. Yet a physical record of the acoustic wave, considered by itself, shows nothing that corresponds to the segmental organization that seems so obvious to a human listener. Moreover, it became obvious after the fact that there could be no segmental organization of the acoustic signal paralleling the phonemic organization because humans can transmit more phonemes per time unit than could be resolved by the ear if they were transmitted as unitary signals. At thirty phonemes per second, the human ear would only hear an uninterpretable buzz (Lieberman 1975).

The relative distances between the formants (or bands of energy) and the slope of the formant transitions are major cues in the perception of speech. The formants, like F1 and F2 in figure 1.2, when presented without transitions, will be perceived as vowels, like [i] and [u]. The formant transitions, however, are considerably less obvious in a physical sense, containing perhaps two hundred times less acoustic energy than the formants, yet they have a disproportionate share of linguistic saliency. Other consonantal cues, like the difference between voiced and unvoiced consonants such as [b] and [p], may also involve small vibratory signals that begin about twenty msec before the onset of the formant. In all of these cases, the cues for consonants are not the most physically salient component of the acoustic signals; yet

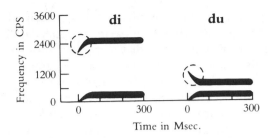

FIGURE 1.2   Simplified spectrographic patterns sufficient to produce the syllables [di] and [du]. From *Cognitive Psychology* 1:301–
323 (1970). *(Used with the permission of Alvin M. Liberman.)*

there is evidence that human neonates of two weeks of age are attending to
the differences between voiced and unvoiced consonants. This can be
shown by experiments using machine-synthesized speech (Siequeland and
DeLucia 1969; Eimas et al. 1971; Cutting and Eimas 1975).

Human infants spend a good deal of time engaged in nonnutritive
sucking, and they will readily suck on an artificial nipple that is wired to a
polygraph. The polygraph will produce a continuous record of the nonnutritive sucking, including any pauses the infant might make. Infants will
interrupt their sucking behavior to attend to novel stimuli, and this propensity can be used to test the novelty value of different speech sounds. By
using artifically produced phonemes, it is possible to repeat one speech
sound, say /p/, until the infant loses interest and returns to nonnutritive
sucking. At this point the cue is changed slightly each time it is presented
until it reaches the critical point as perceived by adults. If the child hears a
change, it will respond with attention to the novel stimulus, reflected in a
pause in the sucking. Using this procedure, it is possible to show that infants
can hear one difference between voiced and unvoiced stop consonants because they interrupt their sucking when the voicing cue straddles the twenty
msec boundary.

As several authors have pointed out, there are some striking parallels
between the instinctive systems studied by ethologists and the phonemic
systems studied by psycholinguists (Marler 1970, 1975; Mattingly 1972). In
the *sign stimuli* discovered by ethologists in animals, such as the spot on the
beak of a herring gull that releases gaping in the chick, the signal is innately
salient and continues to provoke a reaction in an observer even if removed
from a natural context or simulated by an artificial model (Tinbergen and
Perdeck 1950). Using the pattern-playback, it is possible to construct artificial phonemes which preserve only a small part of the acoustic information
contained in natural speech. Such simulations have a hollow, unnatural
sound, since they contain no idiosyncratic information about the speaker,
but they are nonetheless perceived as speech by human subjects. Even more

important, the most salient sign stimuli are not necessarily the most natural ones. Rather, ethology introduced the term *superstimulus* for those artificially created models of sign stimuli that were more efficacious in inducing effects in animals than the naturally occurring signals. Similarly, in psycholinguistics, artificially produced formant transitions that have steeper slopes than naturally occurring ones will be more consistently labeled than the realistic examples. Furthermore, the fact that human neonates attend to some of these cues and children acquire phonemic systems simply as a function of exposure suggest that they are innately salient. Also, as already mentioned in the context of bird song, there are innately determined critical periods that partially determine the effects of exposure to acoustic information in instinct-learning interlockings. Although no one has reared children in isolation to see if they will speak Hebrew, the Biblical first language, there is a great deal of circumstantial evidence to suggest that the phonemic habits of first languages are much more resistant to subsequent learning than the morphological and lexical components of language and that the modifiability of phonological habits decreases significantly after puberty.[51] Moreover, it is of interest that brain damage to the left hemisphere occurring before the age of eight causes language to lateralize in the right hemisphere; but this does not occur in older children, suggesting an age effect on the consolidation of language in the brain. Although it is a fact that human beings can acquire additional languages even as adults and that languages themselves can incorporate both invented and borrowed elements, there is still a fundamental historical continuity that belies the complete arbitrariness of this process. Even in English, which has the reputation of being utterly shameless in its propensity to borrow words, such that only a minority of its one million or more words can be traced to the Anglo-Saxon of a millennium ago, the one hundred most frequently used words today are from the Old English vocabulary, as are 83 percent of the next most commonly used one hundred. In spite of lexical eclecticism, English is historically continuous at its semantic core (Kučera and Francis 1967).

The fact of historical continuity in languages, which has been obvious to linguists since the beginning of the nineteenth century, has coexisted with the equally obvious fact that there is no necessary relationship between particular sounds and particular meanings nor between particular gene pools and particular languages. Traditional culture theory, which derived from the progressivism of the sciences and the mechanical arts, cannot account for the inherent phonological conservatism of languages, whereas traditional instinct theory would be incompatible with the cultural variability in phonological systems. However, the recent biological research into the instinct-learning interlockings of bird song development helps provide an alternative model that accounts for the evidence. Like white-crowned sparrows that can learn any white-crowned sparrow song but lock into the one they are exposed to as nestlings, humans can learn any human phonological system but lock into the one they are exposed to in infancy. Like white-crowned sparrows that have different songs in different localities, the immature human has to construct a behavioral system out of innately given constraints and environmental information, and this process creates intergenerational variation, which yields differentiation into dialects. As in white-crowned sparrows, the historical continuity of human language can be interrupted at any point by rearing children in different linguistic environments, but the new language immediately locks in and begins anew the process of historical continuity.

If human language, so long considered to be the cultural phenomenon *par excellence,* develops through the interaction of innate and traditional forms of information, then the commonly accepted anthropological view of culture as a level of organization must be critically examined. Contrary to the latent historicism of most anthropological theory — a historicism made all the subtler by being cast in synchronic terms — an ethological perspective suggests that human evolution is only understandable by postulating innate factors and innate-learning interlockings in man. Furthermore, an ethological perspective suggests that culture is not a unitary system but a generic name for the traditional information component of innate-learning interlockings. Consequently, there are as many forms of culture as there are distinct biological systems that interact with traditional information. Some of these systems may be uniquely human, and others may have species-specific aspects, but the majority of human instinctive systems will be similar to homologous systems in nonhuman primates and all will be historically continuous with morphological precursors discernable in contemporary primates. The methodology of human evolution is to establish these homologous behavioral systems, to characterize their species-specific features, and to evaluate the effects of increased neocortical representation upon their activity. This approach explicitly rejects the proposition that human evolution has been characterized by the replacement of instinct by culture; and it also rejects the proposition that culture is a level of organization that can be understood without reference to the psychobiological mechanisms that process traditionally transmitted information. To the contrary, traditional transmission can only be assessed in the context of specific sensory, motivational, and motor mechanisms that are themselves innate.

THE LADDER DESCENDED

The opposition between instinct and learning that modern social science has carried over almost intact from the Enlightenment is of dubious biological validity, but its scientific status was greatly enhanced until the present day by a theory of the nervous system that seemed to support it.[52] The second half of the nineteenth century, the period that gave rise to the academic disciplines of modern behavioral science, also made a number of fundamental discoveries about the nervous system. Even before biological evolution had been suggested, the anatomical homology of the human brain to that of other primates was established and the increased size of the neocortex was demonstrated to be a feature of the primate in general and of man in particular (Müller 1843). Beginning about the 1860s, clinical neurology was able to show that damage to areas of the neocortex in man led to deficits in particular aspects of activity, such as language and learned skills, that were characteristically human. Also, it was shown that language was controlled primarily by a single hemisphere, a phenomenon known as cerebral lateralization, whereas no such effects could be demonstrated in animals. Clinicians also observed that neocortical damage that destroyed propositional speech might still spare automatic and emotional aspects of language, like prayers and swearwords. Comparative anatomy also attempted to show that the neocortex was a mammalian innovation but that the subcortical and "old" cortical regions were homologous with the brains of reptiles. All of these findings were compatible with a progressivist interpretation of neural organization, in which neocortex, associated with learning and the subordination of "lower" instinctive centers, was seen as a later phylogenetic emergent particularly characteristic of man.[53] This body of neural theory forms the background to all twentieth-century discussions of the evolution of the human brain. Physical anthropology has had from the beginning close institutional connections with the medical disciplines of clinical neurology and human anatomy, as such well-known anthropological anatomists as Le Gros Clark (1963), Elliot Smith (1924), and Arthur Keith (1948) bear witness. Moreover, it was the French physician and anatomist, Paul Broca, discoverer of the so-called speech area of the dominant cerebral hemisphere (which today bears his name), who is credited with the founding of the Societé d'Anthropologie in Paris in 1859.[54] This date is usually taken as the beginning of modern anthropology in France, as it was the first anthropological institution to be established after the decline of the Degerando group (which had produced a program for ethnological investigations) two generations earlier; and it led to the subsequent founding by Broca in 1875 of the first teaching facility for the subject in France, the École d'Anthropologie.

The integrated theory of the human brain, formulated by the neuroscientists of the second half of the nineteenth century, also enters modern social thought as the implicit body of assumptions behind the stratified conception of id, ego, and superego in psychoanalysis. Freud was a neurologist by profession, and he in fact had received one of the finest educations in neuroscience that his age could provide. His professor at Vienna, Theodor Meynert, was one of the formulators of the integrated theory of the human brain, and Meynert specifically defined psychiatry as the study of diseases of the forebrain, the progressive component of the nervous system (1885). In other hands, this neuroanatomical theory, reasonable in many respects, came very close to recasting the philosophy of history in neurological terms. Casual students of the nervous system have often transformed it into a caricature, best termed the *Victorian brain,* from which not even Nobel laureates in biology in the year 1977 have proved to be immune.[55]

Human evolutionary studies continue to be hampered by the emphasis that this theory places on the progressive aspects of neural evolution at the expense of continuity. The fact remains that human brains are homologous with the brains of apes, and the known anatomical differences between human and ape brains are morphologically trivial. In fact, there are no parts of a human brain that do not have homologs in the brain of an ape. The human brain is larger — about three times larger than the brain of an ape of comparable body size — and the neocortex occupies a greater fraction of the total brain (Passingham 1973; Jerison 1973; Holloway 1966). Functionally, there are differences between human and ape brains, as behavioral observation alone would suggest; but the most dramatic difference, extensive control of a function by a single side of the brain, occurs in some nonprimate animals, as diverse as birds and sea slugs, and there is evidence to suggest that it can be induced in monkeys under certain experimental conditions (Harnad et al. 1977; Dewson and Burlingame 1975; Nottebohm 1970). In man, lateralization of function is paralleled by a slight size difference between the left and right hemispheres, but similar differences occur in apes, and no new structures appear to be involved (Bonin 1962; Geschwind and Levitsky 1968; Groves 1973; LeMay 1976). So, too, the concept of a "reptilian brain" buried beneath the neocortex of mammals, implied by Victorian theory, has also come to be questioned by neuroanatomy (Masterson et al. 1976a, 1976b). Since evolution is a theory of transformation as well as a theory of continuity, anatomical and functional differences between the brains of different species are to be expected, and doubtless differences between human and ape brains will eventually be discovered. However, such differences are not likely to overshadow the extensive homologies that are already known to exist.

Simply in terms of the comparative evidence already available, a great deal of behavioral continuity between man and animal is to be expected, and the instinctive systems that function in animals can be reasonably expected to have parallels in man. As Darwin himself attempted to show, and as modern cross–cultural research has gone far toward confirming, the emotional communication of man has a large instinctive component and is homologous with that of nonhuman primates.[56] Moreover, as Bateson (1968) has pointed out, this emotional expressive system is not residual in man but is at least as well developed as in apes. Emotional expression is central to all face-to-face interactions in man, and persons rendered incapable of this form of activity, such as those suffering from Parkinson's disease, which can prevent facial display, are not therefore regarded as exemplary human beings, as traditional instinct theory might recommend, but as sick and unfortunate people.[57] These facts suggest that the phylogenetically older emotional brain is no less important in man than in any other mammal, and that a theory of human evolution presupposing the development of reason at the expense of emotion or of learning at the expense of instinct is in serious conflict with the behavioral evidence. So, too, if the Victorian view of the brain were correct, those phylogenetically progressive functions, like language, that can be demonstrated to be mediated largely by neocortical mechanisms, would approximate the affective neutrality of the logicians. But language ordinarily occurs as speech, which is highly sensitive to emotional control and almost invariably associated with those paralinguistic features like volume of voice and rate and pitch of speech that frequently encode the emotional state of the speaker. The speech act too is embedded in the context of social interaction, sustained and modulated by a complex of postural, gestural, and temporal cues that cannot be assumed to be any simple product of the activity of lower centers. The pervasiveness of human emotionality can be noted by the most superficial observer of human behavior, and the similarity of human and animal emotional displays is no less generally accessible. Anthropological theory, however, has chosen until recently to ignore this dimension of human existence in favor of a moral devaluation of instinct and emotion best exemplified by the philosophy of Émile Durkheim.

Darwin's own view of instinct, which he confided to one of his notebooks twenty years before he wrote *On the Origin of Species* is a striking contrast not only to Durkheim's *Elements of Morality* but also to the equally distrustful instinct theory popularized by Freud and psychoanalysis (insertions in brackets are Darwin's own):

ON THE MORAL SENSE

Looking at Man, as a Naturalist would at any other Mammiferous animal, it may be concluded that he has parental, conjugal and social instincts, and perhaps others — The history of every race of man shows this, if we judge him by his habits, as another animal. These instincts consist of a feeling of love [& sympathy] or benevolence to the object in question. Without regarding their origin, we see in other animals they consist in such active sympathy that the individual forgets itself, & aids & defends & acts for others at its own expense. — Moreover any action in accordance to an instinct gives great pleasure, & such actions being prevented by [necessarily] some force give pain: for instance either protecting sheep or hurting them. — Therefore in man we should expect that acts of benevolence towards fellow [feeling] creatures, or of kindness to wife and children would give him pleasure, without any regard to his own interest. Likewise if such actions were prevented by force he would feel pain.[58]

The naturalistic approach to instinct, of which ethology is the modern representative, explicitly denies that human instincts are to be regarded as primitive and residual biological systems that stand in opposition to progressive faculties of reason and morality in which real human nature resides. The naturalistic approach denies too that instincts are any less central to the life of man than they are to any other mammal. In this respect, the instinct theory of Darwin, by accepting a biological component to the moral sense, was far closer to the social theory of Rousseau and the Scottish Enlightenment than to his contemporary John Stuart Mill or to his self-appointed successors, the Social Darwinists.

It was Rousseau, in *The Discourse on the Origins of Inequality among Men,* who had asked rhetorically where man would be, for all his morality, if he were not endowed by nature with compassion for his fellow man? It was a conception of human nature that he drew from the English critique of Hobbes, particularly the criticism by the Third Earl of Shaftesbury, some of whose work had been translated into French by Diderot only seven years before.[59] What has been interpreted by so many commentators as the romantic primitivism of Rousseau is the astute sociological observation that some societies may be better than others in allowing compassion to express itself. It was this point that Adam Smith cited approvingly in his review of the *Discourse of Inequality* and that he later referred to in a far more sophisticated treatment of his own, *The Theory of Moral Sentiments* (Adam Smith 1812, vol. 3; Hartley 1810).[60] So, too, the pleasure attendant on doing well by one's fellow man was also recognized by that most utilitarian of utilitarians, Jeremy Bentham, as a factor to be reckoned with in the calculus of pain and felicity (Bentham 1970, pp. 100ff.). John Stuart Mill, though the godson of Jeremy Bentham, comes closer to the Durkheimian position: following Malthus, he saw the exercise of "prudential restraint" over the sexual impulse as a way for the working classes to circumvent the Iron Law of Wages: "Poverty, like most social evils, exists because men follow their brute instincts without due consideration. . . . Civilization in every one of its aspects is a struggle against the animal instincts" (J. S. Mill 1965, vol. 3, pp. 760–761; J. S. Mill 1869, vol. 2, p. 272).[61] And sympathy, although it has not disappeared in Mill's thought, is given a much more restricted conceptual role. In spite of such qualifications, the psychological theory of the Scottish Enlightenment in recognizing a natural psychological component of society was not necessarily committed to a view of human nature in which socialization was in the saddle and rode mankind. Early utilitarianism, by viewing the social order as the collective product of the clash of individuals pursuing their private interests, and by including fellow feeling and humane regard, at least in theory, as among the innate human motives that entered the calculation of interest, must be regarded as an important precursor of a naturalistic conception of human nature typified by the Darwinian theory of instinct, just as its population theory was a precursor to the mechanism of natural selection itself. However, mainstream social theory, from the middle of the nineteenth century on, has only been too happy to climb the ladder of reason and exercise its own "prudential restraint" over instinct as an explanatory concept of the affairs of men.

MATCHLESS EFFRONTERY

The ethological approach, by emphasizing the genetic continuity of brain mechanisms that underlies behavior, gives nonhuman primates a central place in the theory and practice of Western social thought. However, their significance was impossible to perceive as long as human evolution was construed as a ladder of reason. Even in scientific circles these animals were most commonly thought of as poor parodies of man, and popular language still reflects their inferior status. A behavioral approach to human evolution requires exactly the same process of decentration that makes possible an anthropology of savages, and with a few exceptions, like Charles Darwin and Friedrich Engels, Victorian theorists of human evolution were no more disposed to consider the social organization of monkeys in a self-reflective vein than they were the societies of non-Europeans. As with the beginnings of an ethnological consciousness, a decentered approach to nonhuman primates began in the mideighteenth century, in that burst of rationalist humanism that Kant called the Age of Enlightenment. As with ethnology itself, the *philosophes* did not contribute a new body of facts but began to perceive the historical significance of the few facts then available.[62]

Monkeys were known to Europeans from ancient times, and one species still lives on the margins of the Mediterranean, but the nonhuman primates most similar to man, the anthropoid apes, did not become known to science until late in the seventeenth century. Nicolas Tulp, the anatomist portrayed in Rembrandt's *Anatomy Lesson,* provided a drawing and a firsthand description of an ape that had been brought to Europe, most probably a pygmy chimpanzee. In 1698, an English anatomist, Edward Tyson, published a description of another chimpanzee, and in the eighteenth century, orangutans from the Dutch East Indies were imported into Europe. Linnaeus was the first to classify formally man with the ape in a biological taxonomy, in 1735, in a class he called *Anthropomorpha,* or "human-formed," later changed in the tenth edition of his work to *Primates,* or "first among animals." However, the classification is based upon anatomical, not behavioral, criteria.[63] Linnaeus, in fact, thought that man was so similar to apes anatomically that it was difficult to specify distinguishing characteristics.

Rousseau, in footnote 10 to the *Discourse on Inequality,* was among the first to draw attention to the behavioral significance of apes, in commenting on an account by an English traveler to Angola, Andrew Battell, whose observations of chimpanzees were printed in a collection of voyages, *Hakluytus Posthumus,* or *Purchas His Pilgrimes,* published in 1625, and read by Rousseau in French translation.[64] It cannot be asserted these creatures are animals, Rousseau argues, until it is empirically demonstrated that their primitiveness is due to an absence of the human faculty of "perfectability," equivalent to cultural capacity in Rousseau's usage, and not to an insufficiency of training. Here, for the first time, the "human" status of apes is made an empirical question. Not long after, in 1760, Hoppius, a student of Linnaeus, in the treatise *Anthropomorpha,* reviews the behavior of apes, which he finds very manlike, and, echoing Rousseau, he laments the absence of behavioral primatology.[65] In Linnaeus's own *Systema Naturae,* however, comparative behavior is confined to didactic essays, moralistic in tone, that are appended to the classification. The generalizations are questionable even by eighteenth century standards: "That in an uncultivated stage you [Man] are foolish, lascivious, imitative, ambitious, prodigal, anxious, cunning, austere, envious, avaricious, and get *transformed* so as to be attentive, chaste, considerate, modest, sober, tranquil, sincere, soft, beneficent, content."[66] We can imagine what Rousseau would have said to this, and it is perhaps with some justification that Diderot dismissed Linnaeus as "that Methodist."[67]

The primatological awareness of Rousseau and Hoppius was not generally shared, and the ape was more likely to be drawn as a figure of ridicule. Thus, in Buffon's Eurocentric taxonomy of the animal world, in which order decreases as we move outward from Europe and wildness is the absence of civilized refinement, the monkey is a most embarrassing creation, so like man physically but in his manners the logical opposition of all that Buffon held dear. It perhaps is not surprising that in the *Natural History* man comes at the beginning of the series of descriptions of mammalian species and the baboon at the very end.[68] Buffon's description of the baboon is one that anticipates in its own way what Darwin later called the "astonishing licentiousness of savages" (1901, p. 896):

> He is a squat animal, whose compact body and nervous members indicate strength and ability. He is covered with long close hair, which gives him the appearance of being larger than he is in reality. His strength, however, is so great, that he would easily overcome one or several men, if not provided with arms. Besides, he is continually agitated by that passion which renders the gentlest animals ferocious. He is insolently salacious, affects to show himself in this situation, and seems to gratify his desires *per manum suam,* before the whole world. This detestable action recalls the idea of vice, and renders disgustful the aspect of an animal, which Nature seems to have particularly devoted to such an uncommon species of impudence; for, in all other animals, and even in man, she has covered

these parts with a veil. In the baboon, on the contrary, they are perpetually naked, and the more conspicuous, because the rest of the body is covered with long hair. The buttocks are likewise naked, and of a blood red colour; the testicles are pendulous; the anus is uncovered, and the tail always elevated. He seems to be proud of all those nudities; for he presents his hind parts more frequently than his front, especially when he sees women, before whom he displays an effrontery so matchless, that it can originate from nothing but the most inordinate desire.[69]

In contrast to Buffon, the naturalist's regard for the behavior of apes was most systematically developed in the eighteenth century, perhaps independently of Rousseau, by James Burnet, a Scottish jurist better known as Lord Monboddo. He elaborated the same logic as Rousseau's *Second Discourse* into six volumes published between 1774 and 1792, of which volume 1 is the most interesting to a modern evolutionist. Monboddo's thesis stands in direct opposition to Buffon's passage on man, quoted previously, which defines man as a species with the body of an ape and the faculties of reason and speech.

Although Monboddo believed in essential natures as much as anyone else of his time and developed his thesis within an explicitly metaphysical framework, he also had a truly historical consciousness, in which not even the defining features of man are immutable and fixed but show a development to their present form. It was this sense of history that led Monboddo to ask whether language and thought were natural to man or whether they were acquired by man as a member of society.

One of his major lines of evidence is the orangutan, which at that time was a generic name for anthropoid apes, with not much care taken to distinguish the Asian orangutan carried home by the Dutch from the African chimpanzee described by Tulp and Tyson and reported on by Battell. Monboddo concluded from these accounts that orangutans were in fact a species of man that had not yet acquired the faculty of speech, and he was buttressed in this interpretation by Buffon's own assertion (an incorrect one) that the anatomy of articulation was the same in apes and men. If we observed the deaf and mute in similar circumstances, he argues, would we not conclude that they were beasts?

> With such philosophers; it would be in vain to argue, that, having the human intelligence, and likewise the organs of pronunciation, they must, necessarily have the capacity of learning, by teaching and imitation, if not of inventing a language; and, if, he have the capacity of learning to speak, that is sufficient to denominate him a man, though he never attain to the actual exercise of the faculty; because human nature, as we have elsewhere observed, consists chiefly of capabilities. But I say to these gentlemen, *first,* that the experiment has never been fairly tried upon any Orang Outang that has been

hitherto brought to Europe. For it does not appear that any pains were ever taken to teach any of them to speak. We cannot therefore affirm that they would not learn the art, if the same pains were to be bestowed upon them that Mr. Braidwood bestows upon his scholars [deaf children]. But, *secondly*, I say, that, if the experiment should not succeed, it would not prove that the Orang Outang is not a man. For the habits and dispositions of mind, and, by consequence, the aptitude to learn any thing, are qualities which go to the race, as well as the shape and other bodily qualities. And it is for this reason, that the offspring of a savage animal will never be so tame, whatever pains may be taken upon him, as the offspring of a tame animal. And, I am persuaded, it is with wild men, as with wild fruits, which we know will not lose their savage nature at the first remove, but can only be tamed by continued culture for a succession of generations. (Monboddo 1774, 1:299–300)

If Rousseau complained his contemporaries failed to understand the *Discourse on Inequality*, he had at least numerous intellectual progeny who thought they did; but Monboddo cannot be called an ancestor because he had no descendants. His contemporary, J. R. Forster, sets aside three pages of *Observations* to counter the argument that orangutans are of the same species as man. However, he totally neglects the interesting part of Monboddo's argument, the historical perspective; and he finally goes on to ridicule orangutans as ugly (J. R. Forster 1778, pp. 253– 256). Adam Ferguson, a few years before, could similarly dismiss Rousseau by failing to see the state of nature in any but synchronic terms (1773, pp. 8– 9):

> In opposition to what has dropped from the pens of eminent writers, we are obliged to observe that men have always appeared among animals a distinct and superior race; that neither the possession of similar organs, nor the approximate of shape, nor the use of the hand, nor the continued intercourse with this sovereign artist, has enabled any other species to blend their nature or their inventions with his; that in his rudest state, he is found to be above them; and in his greatest degeneracy, never descends to their level. He is, in short, a man in every condition; and we can learn nothing of his nature from the analogy of other animals.

Monboddo, considering "examples from ancient and modern History of Men living in the Brutish State, without arts or civility" (1774, vol. 1, bk. 2, chap. 3, pp. 236– 269), came to the conclusion that reports of tailed men should not be dismissed as impossible. In this conclusion he was joined by Linnaeus, and also Hoppius, who illustrated a tailed man in the *Anthropomorpha*.[70] Monboddo argued that the os coccyx is anatomically a tail and that physicians have observed works of nature even odder, such as children born with six fingers. Such a dissertation offended common sense, and it provoked Samuel Johnson, a man of uncommon common sense, to give voice to the general opinion in one of his least discerning jibes: "Other people have strange notions," he reflected, "but they conceal them. If they have tails, they hide them; but Monboddo is as jealous of his tail as a squirrel."[71]

NOTES TO CHAPTER 1

1. See Vaughn's (1962) Introduction to Rousseau's political writings; also *Confessions*, pp. 361–363. I have used the 1817 French edition of the *Discourse on Inequality* (Rousseau 1817, Tome 3) and Masters's translation (Rousseau 1964).

2. Adam Smith's review is in Smith (1967, pp. 23–28). See also Edmund Burke (1756) for a favorable evaluation. Johnson's remarks are in Boswell (1960, p. 405: said on 30 September 1769).

3. Rousseau is not mentioned in Robert Lowie's *History of Ethnological Theory*, nor in the first four chapters of Fortes's *Kinship and the Social Order*, where the history of social anthropology is reviewed. Marvin Harris gives Rousseau five sentences in the 806 pages of *The Rise of Anthropological Theory*, where his work is placed among "those romantic mystifications of history which replace the notion of natural law with unpredictable and ungovernable national or tribal collective souls" (pp. 21–23); and Penniman, in his one paragraph, tells us that Rousseau was the first to introduce the noble savage to the world's literature (*A Hundred Years of Anthropology*, p. 51.).

4. *The Confessions* (Rousseau 1975, p. 362). See also Lovejoy (1923).

5. Hewes (1977) gives priority to La Mettrie (Vartanian 1960) for raising the possibility of ape language. However, La Mettrie may simply have priority of publication, since much of his thought is apparently derivative (Roger 1971). Rousseau's remarks are contained in n. 10 to the *Discourse on Inequality*, which, in another of those lapses of scholarship that seem to have dogged Rousseau's migration into English, is missing from a widely used English translation (Rousseau 1916). The relevant section, also given in Masters's translation (Rousseau 1964), reads:

> I am convinced, that in the three or four hundred years that the inhabitants of Europe have been swarming over the rest of the world, and continually publishing new collections of travel stories and journals, the only men we take cognisance of are Europeans. Furthermore, from the ridiculous prejudices that are still prevalent even among men of letters, it seems that under the pompous name of the study of man, each one studies only the men of his own country. Private individuals may well come and go; philosophy does not seem to travel at all, and so each people's philosophy is inappropriate for another. The reason for this is obvious, at least as far as distant countries are concerned. There are really only four kinds of men who make long voyages: sailors, merchants, soldiers and missionaries. Now, one would hardly expect the first three to make good observers; as for the last, (even though they are not subject to the same preconceptions as others) one must believe that they would not willingly give themselves up to pursuits which seem to be idle curiosity, and which would divert them from the more important works to which they are destined, preoccupied as they are with the sublime vocation that has called them. Moreover, to preach the gospel effectively, one only needs zeal and God gives the rest; but to study man, one needs some talent which God promises no one, and which is not always in the allotment of saints. One cannot open a travel book without finding descriptions of characters and customs; but one is quite surprised to see that these people who have described so many things have only said what everyone already knows; at the

other end of the world, they only took note of what they would have been able to see without leaving their own street. The real characteristics which distinguish nations, which catch the trained eye, have almost always escaped them. From this has come that pretty moral adage, so bantered about by the pseudo-philosophical mob: that men everywhere are the same and that since vices and virtues everywhere are the same, it is rather pointless to try to characterise different peoples. This argument is just about as well thought out as if one were to say that one cannot tell the difference between Peter and Paul because they both have a nose, a mouth and eyes.

One admires the liberality of those curious few who have made or commissioned costly trips to the East with scholars and painters to draw hovels and to decipher or copy inscriptions. However, in an age that prides itself on its fine attainments, I find it difficult to understand that one cannot find two men in accord, both rich, one moneyed, the other talented, both loving honour and aspiring to immortality, one of whom will give up 20,000 ecus of his fortune, and the other ten years of his life, to a famous voyage around the world, where they would study not just monuments and plants, but for once men and customs; men who, after so many centuries spent measuring and contemplating the house, at last take it upon themselves to want to know its inhabitants.

(Trans. Hilary Yerbury and Peter C. Reynolds.)

6. See Flavell for decentration (1963, pp. 230–231) and for egocentricity (pp. 60, 64–65, 224, 279, and the references cited therein).

7. Lubbock's incapacity for decentration provoked Karl Marx (1974, p. 340) to comment: "These civilized asses!" Burrow (1966, pp. 228–229) discusses Lubbock's relationship to Darwin.

8. Contemporaneous accounts of early archeology can be found in Lubbock (1869) and Lyell (1873). Daniel (1950, 1976) presents a historical account of the development of the discipline. Though the interpretation of the archeological record in terms of stages was prominent in the later nineteenth century (Daniel, 1950, p. 244), field archeology was always primarily concerned with actual history. Mitchell, in 1880 (pp. 17, 23–24, 206–207), was already criticizing the ethnological comparative method as used in archeology. The history of the study of early man and geology can be found in Geikie (1897) and Green (1959) among others. Linguistics is discussed below (this chap). Collingwood (1946) treats the idea of history generally.

9. On Buffon see Fellows and Milliken (1972) and Wilkie (1959). For the evolutionism of Maupertuis, see Glass (1947), Maupertuis (1974) and Diderot in *D'Alembert's Dream* (Diderot 1966). Diderot's essay, written in 1769, is much in advance of early nineteenth century conceptions, but it was not published until after the Napoleonic wars. A detailed look at eighteenth century biology is provided by Roger (1971).

10. Buffon's *Natural History* went through fifty-two complete editions in French, eight in German, six in Italian, four in English, one in Spanish, one in Dutch, and more than 325 partial editions (Fellows and Milliken, 1972, p. 16). For Rousseau's enthusiasm for Buffon, see Fellows (1960).

11. Buffon, *Natural History*, article on the ass, from Barr's translation (1797, vol 5. pp. 182–183).

12. For the historical development of anatomy, with early illustrations of homological comparison, see F. J. Cole (1944).

13. For the philosophical background to the eighteenth century, see Cassirer (1951) and Gay (1970).

14. D'Alembert compares Locke to Newton (Cassirer 1951, p. 99).

15. For early biochemistry, see Fruton (1972) and Leicester (1974), especially chap. 13 of the latter.

16. The quote is from a review of Soame Jenyns's book, *Review of a Free Inquiry into the Nature and Origin of Evil* (1757). Quoted by Lovejoy (1936, p. 254).

17. The work of Smellie, one of Buffon's English translators, is abstracted in R. Kerr (1811) and also summarized by Bryson (1945, p. 62).

18. Quoted by Lovejoy (1936, p. 279) from *De la Nature,* vol. 4, pp. 17–18.

19. Kant, in Hastie (1900, p. 145). See also Popper (1963, pp. 175–183), for discussion of Kant's cosmogony.

20. See Kant (1950) and Scheidt (1950) for Kant's racial theory.

21. Jone's contribution to linguistics is discussed in this chap. The quotation is given by Edgerton (1946).

22. Ethnology, defined as the intent to record the ways of life of non-European peoples in order to establish a science of man, really begins with the voyages of exploration in the mideighteenth century, particularly to the Pacific. James Douglas, president of the Royal Society, formulated a "notes and queries" for anthropology prior to the first Cook voyage, in 1768 (Beaglehole, 1955, vol. 1, app. 2). The more elaborate and better known Dégerando memo (1969) was presented later (1800).

23. With the exception of Commerson's letter (1769) on Tahiti, the noble savage is conspicuously missing from the journals of the voyages in the Pacific, although explorer's accounts were often subsequently ennobled by commentators and artists back in Europe (Parks 1964; B. Smith 1960). Keen's (1971) detailed examination of European accounts of the Aztecs, from the time of their discovery to the twentieth century, reveals that very little good was ever said about them. Moreover, the concept of the noble primitive has ancient roots in Western thought (Cohen 1957; Lovejoy and Boas 1935).

24. Montaigne (1967), Montesquieu (1964), Diderot's *Reflections on the Voyage of Bougainville* (1966).

25. Meek (1973, 1976) has examined the use of the stage concept of history in the late eighteenth century. See also Condorcet (1955).

26. From Manuel (1965, pp. 110–111). See also Teggart (1960, especially chap. 10) and Comte himself on the laws of succession (1880, vol. 2, p. 85), progress (1892, p. 30), and the organic analogy (1892, pp. 263ff.)

27. Letter to Edward Lott, 23 April 1852). In Duncan (1911, p. 62).

28. For example, in *Animaux sans Vertèbres* (1815) Lamarck writes:

> Who does not know that, in its actual state of organization, each living body of what kind soever is a really perfect being, that is, a being which lacks nothing of that which is necessary for it! But nature having complicated animal organization more and more, and thereby having been enabled to endow the animals having the most complex organization with more and more advanced abilities, it is possible to see in the last term of her efforts a perfection from which those animals which have not attained it can be seen to recede in stages. (Lamarck 1914, pp. 58–59; see also p. 68).

29. Cf. Spencer's definition of structure and function (1885, pp. 473ff.) and his summary of his theory (1885, pp. 579–585). For Spencer's connection to Durkheim, see Durkheim (1964*b*) and Lukes (1972).

30. The history of Victorian anthropology can be found in Burrow (1966), Haller (1970, 1971) and Stocking (1968, 1971). Keen's (1971) chapter on Morgan is especially revealing.

31. Social Darwinism is recounted by Hofstadter (1945) and Burrow (1966, p. 20). Engels writing in the 1870s must be given both the first and last word on this subject: "Until Darwin, what was stressed by his present adherents was precisely the harmonious cooperative working

of organic nature. . . . Hardly was Darwin recognized before these same people saw everywhere nothing but *struggle"* (Engels 1941, p. 208).

32. Green (1959, p. 300) quotes Wallace's letter to Darwin.

33. On Darwin on higher and lower, see F. Darwin (1903, vol. 1, p. 114).

34. Engels (1941) gives a scattered treatment of Darwin in *Dialectics of Nature,* recognizing his dialectic approach to history but disliking his Malthusianism (pp. 18–20, 208–210, 235–240). Elsewhere he defends Darwin from Dühring's attack (*Anti-Dühring,* n.d., pp. 78–87).

35. Article on the nomenclature of apes, Smellie translation (1866, vol. 2, p. 43).

36. On the Monogenist triumph, see Haller (1970).

37. For background on Forster, see Hoare (1976).

38. Hume's remarks are in the footnote in the 1753–1754 edition of the essay "Of National Characters," first published 1748. Quoted by Jordan (1969, p. 253).

39. Data in Jerison (1973) show that hominid fossils fall between existing nonhuman primate species and man.

40. Kuper (1973) has written a history of "British social anthropology." See also Rivers (1926, chap. 1, a reprint of a 1916 paper; and 1924). Also Fortes (1969) summarizes some of this period. Lowie (1914) led a similar reaction in America.

41. See p. 181 (Radcliffe-Brown 1952) for a comparison between organisms and society that illustrates the typological cast of his thinking. Comte's original is worth quoting: "The object of science is to discover the laws which govern this continuity, and the aggregate of which determines the course of human development. In short, social dynamics studies the laws of succession, while social statics inquires into those of coexistence. . . ." (1880, vol. 2, p. 84).

42. If Frank Beach didn't say this, he should have. In any event, he said something similar (Beach 1950). See also Hilgard (1960) and Woodworth (1952).

43. For the assumptions made by psychoanalysis in the theory of the id, see Yankelovitch and Barrett (1970, p. 386, and chap. 3).

44. Sapir (1907–1908) gives a review of this work of interest to anthropologists. It is noteworthy too that Herder later climbed the ladder himself (n.d., bk. 5, chap. 6), but here he is arguing for a unitary concept of intellect.

45. This section heading is the title of a chapter in Manuel (1959), which discusses the German reaction to French rationalism.

46. I have used Edgerton (1946) for William Jones and Rocher (1968) for Hamilton. An English version of one of Bopp's early treatises is available (Bopp 1974), with a useful essay by Guigniaut (1877). Rask (1976) is discussed by Malone (1952), Pedersen (1931), and Hjelmslev (1950–1951). Further linguistic background is in Delbrück (1976), Mosse (1963), Pederson (1931), Verbug (1950), and Schlegel (1852).

47. Quoted by Edgerton (1946).

48. Hermes Trismegistus is discussed in the context of Copernicus and Bruno by Yates (1964). An earlier example is the scholarly treatment of the Donation of Constantine (Toulmin and Goodfield, 1965, pp. 104–106).

49. This is in Banks's journal (Beaglehole 1962, p. 371). Banks and Jones moved in the same scientific and scholarly circles in London between the time Banks returned from the Pacific and the time Jones left for India, a period of nearly fifteen years.

50. Linnaeus also inspired Joseph Banks (Beaglehole 1962, p. 18), as well as a whole generation of scientific field workers. His role in the foundation of the field work tradition is central (cf. Rauschenberg 1968, pp. 10–11 and map and illustrations in Blunt 1971, p. 186).

51. Lenneberg (1967) has summarized much of these data.

52. For the background to nineteenth century brain research see Gould (1978), Young (1970), and Magoun (1960). William James (1890) presents a contemporary review.

53. Meynert (1885) and Morgan (1913) elaborate on this position.

54. For the development of French anthropology, see Bender (1965).

55. Jacob (1977, p. 1166) writes:

> To the old rhinencephalon of lower mammals a neocortex was added that rapidly, perhaps too rapidly, took a most important role in the evolutionary sequence leading to man. For some neurobiologists, especially MacLean, these two types of structures correspond to two types of functions but have not been completely coordinated or hierarchized. The recent one, the neocortex, controls intellectual, cognitive activity. The old one, derived from the rhinencephalon, controls emotional and visceral activities. In contrast to the former, the latter does not seem to possess any power of specific discrimination, or any capacity for symbolization, language, or self-consciousness. The old structure which, in lower mammals, was in total command has been relegated to the department of emotions. In man, it constitutes what Mac-Lean calls "the visceral brain." Perhaps because development is so prolonged and maturity so delayed in man, these centers maintain strong connections with lower autonomic centers and continue to coordinate such fundamental drives as obtaining food, hunting for a sexual partner, or reacting to an enemy. This evolutionary procedure — the formation of a dominating neocortex coupled with the persistence of a nervous and hormonal system partially, but not totally under the rule of the neocortex — strongly resembles the tinkerer's procedure.

MacLean's paper (1949) certainly lends itself to the Victorian brain interpretation, even if the author skirts the line.

56. See chap. 3.

57. See chap. 3.

58. C. Darwin, in "Old and Useless Notes," no. 42, 5 May 1839. In Gruber and Barrett (1974, p. 398.)

59. Diderot translated Shaftesbury in 1745 (Hendel 1934, vol. 1, p. 39). Rousseau and Diderot were still good friends at this stage of their lives.

60. For further sources on associationist psychology, see Mill (1869), Mill (1969), Young (1970), and Spencer (1870).

61. On J. S. Mill's godson status, see Mazlish (1975, p. 75). The quote is from Mill (1965, vol. 2, p. 367). Malthus (n.d.) had presented a similar argument in *Essay on Population*.

62. A historical account of early primatology is in Reynolds (1967, chaps. 1–2); extensive quotations of early accounts are in Monboddo (1774, vol. 1, bk. 2, chap. 4) and in Buffon's article on the orangutan in his *Natural History* (1797, vol. 9).

63. Some of Linnaeus's writings are translated in Bendyshe (1865), with bibliographical remarks (pp. 421–458).

64. Rousseau, *Discourse on Inequality* (1817, n. 10, p. 319). Rousseau adopted Turgot's term "perfectability" and gave it a more naturalistic interpretation. Meek (1976) discusses Turgot's usage. See Hendel (1934, vol. 1) on Rousseau's interpretation of the word.

65. Hoppius's work is partially translated in Bendyshe (1865, pp. 447ff.).

66. Linnaeus quoted in Bendyshe (1865, p. 428).

67. Diderot's attribution is cited in Wilson (1957, p. 194).

68. Jordan (1969) deals with Buffon's racial theories and taxonomy. Buffon's low opinion of New World animals was countered specifically by Thomas Jefferson, who collected statistics to prove that American animals were as big or bigger than European species (Jefferson, 1955, pp. 43–72).

69. Smellie translation (1866, vol. 2, pp. 56–57).

70. Monboddo (1774, vol. 1, pp. 259–263) cites Linnaeus. See also Bendyshe (1865) and the plates in Reynolds (1967) for reports of tailed men. Hexadactylism had been used by Maupertuis to trace the familial transmission of inherited characteristics (Dunn 1965; p. 35; Glass 1947).

71. Quoted by Lovejoy (1933, p. 282).

REFERENCES

ATZ, JAMES W. 1970. The application of the idea of homology to behavior. In *Development and evolution of behavior,* ed. Aronson et al., pp. 53–74. San Francisco: Freeman.

BACHOFEN, J. J. 1967. *Myth, religion and mother right: selected writings of J. J. Bachofen.* Trans. Ralph Manheim. Bollingen Series. Princeton: Princeton University Press; London: Routledge and Kegan Paul.

BARLOW, NORA 1958. *The autobiography of Charles Darwin, 1809–1882,* with original omissions restored. London: Collins.

BATESON, GREGORY 1968. Redundancy and coding. In *Animal communication: techniques of study and results of research,* ed. Thomas A. Sebeok, pp. 614–626. Bloomington: Indiana U. Press.

BEACH, FRANK A. 1950. The snark was a boojum. *American Psychologist* 5: 115–124. Reprinted in *Readings in Animal Behavior,* ed. Thomas E. McGill, pp. 3–15. New York: Holt, Rinehart and Winston.

BEAGLEHOLE, J. C., ed. 1955. *The journals of Captain James Cook on his voyages of discovery. I. The Voyage of the* Endeavour, *1768–1771.* Introduction, pp. xxi–cclxxxiv. Cambridge: Cambridge U. Press.

1962. *The* Endeavour *journal of Joseph Banks, 1768–1771.* Sydney: Angus and Robertson.

BENDER, DONALD 1965. The development of French anthropology. *J. History of the Behavioral Sciences* 1:139–151.

BENDYSHE, T. 1865. The history of anthropology. *Memoirs of the Anthropological Society of London* 1:335–458. London: Turner.

BENTHAM, JEREMY 1970. *An introduction to the principles of morals and legislation.* Ed. J. H. Burns and H.L.A. Hart. London: Athlone Press, U. of London.

BLAKEMORE, COLIN 1974. Developmental factors in the formation of feature extracting neurons. In *The neurosciences: third study program,* ed. F. O. Schmitt and F. G. Worden, pp. 105–113. Cambridge, Mass.: MIT Press.

BLOOMFIELD, LEONARD 1962. *Language.* New York: Holt, Rinehart, and Winston. 1st ed., 1933.

BLUNT, WILFRID 1971. *The compleat naturalist: a life of Linnaeus.* London: Collins.

BONIN, GERHARDT VON   1962. Anatomical asymmetries of the cerebral hemispheres. In *Interhemispheric relations and cerebral dominance,* ed. V. Mountcastle, pp. 1–6. Baltimore: Johns Hopkins Press.

BOPP, FRANZ   1974. Analytic comparison of the Sanskrit, Greek, Latin and Teutonic languages, shewing the original identity of their grammatical structure. In *Amsterdam studies in the theory and history of linguistic sciences,* vol. 3, ed. E. F. K. Koerner. Amsterdam: John Benjamins.

BOSWELL, JAMES   1960. *Life of Johnson.* Reprint of 3rd ed., 1799. London: Oxford University Press.

BOURNE, GEOFFREY H., ed.   1977. *Progress in ape research.* New York and London: Academic Press.

BRYSON, GLADYS   1945. *Man and society: the Scottish inquiry in the eighteenth century.* Princeton, N.J.: Princeton U. Press.

BUFFON, COUNT (GEORGES LOUIS LECLERC)   1797. *Natural history.* Trans. Barr. 15 vols. London: H. D. Symonds.
1866. *Natural history.* 2 vols. (Abridged.) Trans. William Smellie. London: Thomas Kelly.

BURKE, EDMUND   1756. A vindication of natural society, or, a view of the miseries and evils arising to mankind from every species of artificial society. In *The works of the Right Honourable Edmund Burke,* vol. 1, pp. 9–66. London: John C. Nimmo, 1887.

BURROW, J. W.   1966. *Evolution and society: a study in Victorian social theory.* Cambridge: Cambridge U. Press.

BUTLER, SAMUEL   1879. *Evolution, old and new.* London: Hardwicke and Bogne.

CARPENTER, C. R.   1964. *Naturalistic behavior of nonhuman primates.* University Park: Pennsylvania State U. Press.

CARTER, G. S.   1957. *A hundred years of evolution.* N.p.: Sidgwick and Jackson.

CASSIRER, ERNST   1951. *The philosophy of the Enlightenment.* Trans. Fritz A. Koelln and James P. Pettegrove. Princeton, N.J.: Princeton U. Press. German ed. 1932.

CHOMSKY, NOAM   1959. Review of B. F. Skinner's *Verbal behavior. Language* 35:26–58.
1968. *Language and mind.* New York: Harcourt, Brace and World.

COHN, NORMAN   1957. *The pursuit of the millennium: revolutionary millenarians and mystical anarchists of the Middle Ages.* London: Secker and Warburg.

COLE, F. J.   1944. *A history of comparative anatomy, from Aristotle to the eighteenth century.* London: Macmillan.

COLLINGWOOD, R. G.   1945. *The idea of nature.* Oxford: Clarendon Press.
1946. *The idea of history.* Oxford: Clarendon Press.

COMMERSON, PHILIBERT DE   1769. Lettre sur la decouverte de la nouvelle isle de Cythere ou Taïti. *Mercure de France,* Nov. Reprinted in *The quest and occupation of Tahiti by emissaries of Spain during the years 1772–1776,* ed. Bolton Glanvill Corney, vol. 2, pp. 461–466. London: Hakluyt Society, 1915.

COMTE, AUGUSTE   1880. *The positive philosophy.* Trans. and condensed Harriet Martineau. 2 vols. London: John Chapman. 1st French ed., 1853.

CONDORCET, ANTOINE-NICOLAS   1955. *Sketch for a historical picture of the progress of the human mind.* Trans. June Barraclough. London: Weidenfeld and Nicolson. 1st French ed., 1795.

CUTTING, JAMES E., and EIMAS, PETER D.   1975. Phonetic feature analyzers and the processing of speech in infants. In *The role of speech in language,* ed. James F. Kavanagh and James E. Cutting, pp. 127–148. Cambridge, Mass.: MIT Press.

CUVIER, M.   1813. *Essay on the theory of the earth.* Trans. Robert Kerr. Edinburgh: William Blackwood. Facsimile ed. Westmead, Farnborough, Hants., England: Gregg International, 1971.

DANIEL, GLYN E.   1950. *A hundred years of archaeology.* London: Duckworth.

1976. *A hundred and fifty years of archaeology.* Cambridge: Cambridge U. Press.

DARWIN, CHARLES   1966. *On the origin of species, by means of natural selection, or the preservation of favoured races in the struggle for life.* Cambridge, Mass.: Harvard U. Press. Facsimile of 1859 ed.

1862. *On the various contrivances by which British and foreign orchids are fertilized by insects.* London: Murray. 2nd ed. rev., 1877.

1901. *The descent of man and selection in relation to sex.* 4th ed. London. Murray. 1st ed., 1871.

DARWIN, FRANCIS, ed.   1903. *More letters of Charles Darwin: a record of his work in a series of hitherto unpublished letters.* 2 vols. London: Murray.

DÉGERANDO, JOSEPH-MARIE   1969. *The observation of savage peoples.* Trans. by F.C.T. Moore. Berkeley and Los Angeles: U. of California Press. First presented at Société des Observateurs de l'Homme, Paris, 1800.

DELBRÜCK, BERTHOLD   1974. *Introduction to the study of language: a critical survey of the history and methods of comparative philosophy of Indo-European languages.* E. F. K. Koerner, ed. Amsterdam Studies in the Theory and History of Linguistic Science, Amsterdam Classics in Linguistics 8. Amsterdam: John Benjamins. Reprint of 1882 ed., trans. Eva Channing; German ed., 1880.

DENES, PETER B., and PINSON, ELLIOT N.   1963. *The speech chain.* N.p.: Bell Telephone Laboratories.

DEWEY, JOHN   1910. *The influence of Darwin on philosophy and other essays in contemporary thought.* New York: Henry Holt.

DEWSON, JAMES H., III.   1978. Preliminary evidence of hemispheric asymmetry of auditory function in monkeys. In *Lateralization in the nervous system,* ed. S. Harnad. New York: Academic Press.

DEWSON, JAMES H., III, and BURLINGAME, ALLEN C. 1975. Auditory discrimination and recall in monkeys. *Science* 187:267–268.

DIDEROT, DENIS 1966. *Diderot's selected writings.* ed. Lester G. Crocker. Trans. Derek Coltman. New York: Macmillan.

DUNCAN, DAVID, ed. 1911. *Life and letters of Herbert Spencer.* London: Williams and Norgate, 1st ed., 1908.

DUNN, L. C. 1965. *A short history of genetics.* New York: McGraw-Hill.

DURKHEIM, ÉMILE 1914. Le dualisme de la nature humaine et ses conditions sociales. *Scientia* 15:206–221. In *Émile Durkheim, 1858–1917: a collection of essays, with translations and a bibliography,* ed. Kurt H. Wolff, pp. 325–340. Columbus: Ohio State U. Press, 1960.

1961. *Moral education: a study in the theory and application of the sociology of education.* Trans. Everett K. Wilson and Herman Schnurer. New York: Free Press of Glencoe.

1964a. *The rules of sociological method.* 8th ed. Trans. John H. Mueller. New York: Free Press. 1st French edition, 1895.

1964b. *The division of labor in society.* Trans. George Simpson. New York: Free Press.

EDGERTON, FRANKLIN 1946. Sir William Jones, 1746–1794. *Journal of the Oriental Society* 66:230–239. Reprinted in *Portraits of linguists: a biographical source book for the history of Western linguistics, 1746–1963,* vol. 1, ed. T. A. Sebeok, pp. 1–18. Bloomington: U. of Indiana Press, 1966.

EIMAS, PETER D.; SIQUELAND, EINAR R.; JASCZYK, PETER; and VIGORITO, JAMES 1971. Speech perception in infants. *Science* 171:303–306.

ENGELS, FRIEDRICH 1878. *(Anti-Dühring) Herr Eugen Dühring's revolution in science.* Trans. Emile Burns. London: Martin Lawrence. 1st ed., 1878.

1941. *Dialectics of nature.* Trans. and ed. Clemens Dutt. Preface by J. B. S. Haldane. London: Lawrence and Wishart.

FELLOWS, OTIS 1960. Buffon and Rousseau: aspects of a relationship. *PMLA* 75, no. 3 (June): 184–196.

FELLOWS, OTIS E., and MILLIKEN, STEPHEN F. 1972. *Buffon.* New York: Twayne.

FERGUSON, ADAM 1969. *An essay on the history of civil society.* Reprint of 4th ed. Franborough, Hanst. England: Gregg International. 1st ed., 1767.

FLAVELL, JOHN H. 1963. *The developmental psychology of Jean Piaget.* New York: Van Nostrand Reinhold.

FORSTER, JOHANN REINHOLD 1778. *Observations made during a voyage round the world on physical geography, natural history, and ethic philosophy. Part 6: The human species.* London: G. Robinson.

FORTES, MEYER 1969. *Kinship and the social order: The legacy of Lewis Henry Morgan.* London: Routledge and Kegan Paul.

FREEMAN, DEREK   1974. The evolutionary theories of Charles Darwin and Herbert Spencer. *Current Anthropology* 15:211–237.

FREUD, SIGMUND   1962. *Civilization and its discontents.* Trans. James Strachey. New York: Norton. 1st German ed., 1930.

FRUTON, JOSEPH S.   1972. *Molecules and life: historical essays on the interplay of chemistry and biology.* New York: Wiley-Interscience.

GAY, PETER   1970. *The Enlightenment: an interpretation.* Vol. 2, *The science of freedom.* London: Weidenfeld and Nicholson.

GEERTZ, CLIFFORD   1975. The impact of the concept of culture on the concept of man. In *The interpretation of culture,* pp. 33–54. London: Hutchinson.

GEIKIE, SIR ARCHIBALD   1897. *The founders of geology.* London: Macmillan.

GESCHWIND, NORMAN, and LEVITSKY, WALTER   1968. Human brain: left-right asymmetries in temporal speech region. *Science* 161: 186–187.

GHISELIN, M. T.   1969. *The triumph of the Darwinian method.* Berkeley and Los Angeles: U. of California Press.

   1976. The nomenclature of correspondence: a new look at "homology" and "analogy." In *Evolution, brain, and behavior: persistent problems,* ed. R. M. Masterton, W. Hodes, and H. Jerison, pp. 129–142. Hillsdale, N.J.: Erlbaum.

GLASS, H. BENTLEY   1947. Maupertuis and the beginning of genetics. *Quarterly Review of Biology* 22:196–210.

GOLDENWEISER, ALEXANDER A.   1935. *Early civilization: an introduction to anthropology.* New York: F. S. Crofts.

GOULD, STEPHEN JAY   1978. Morton's ranking of races by cranial capacity. *Science* 200:503–509.

GRAY, ASA   1963. *Darwiniana: essays and reviews pertaining to Darwinism.* Cambridge: Harvard U. Press. 1st ed., 1876.

GREEN, JOHN C.   1959. *The death of Adam: evolution and its impact on Western thought.* Ames: Iowa State U. Press.

GROVES, C. P., and HUMPHREY, N. K.   1973. Asymmetry in gorilla skulls: evidence of lateralized brain function. *Nature* 244:53–54.

GRUBER, HOWARD E., and BARRETT, PAUL H.   1974. *Darwin on man.* New York: Dutton.

GUIGNIAUT, M.   1877. Notice historique sur la vie et les travaux de M. Francois Bopp. *Mémoires de l'Academie des Belles-Lettres* 29, no. 1: 201–224. Reprinted in *Analytical comparison. . . ;* pp. xv–xxxviii. Amsterdam Classics in Linguistics 3, 1974.

GUNDERSON, KEITH   1964. Descartes, La Mettrie, language and machines. *Philosophy* 39:193–222.

HALLER, JOHN S., Jr. 1970. The species problem: nineteenth-century concepts of racial inferiority in the origin of man controversy. *American Anthropologist* 72:1319–1329.

1971. Race and the concept of progress in nineteenth century American ethnology. *American Anthropologist* 73:710–724.

HALLOWELL, A. IRVING 1965. The history of anthropology as an anthropological problem. *J. history of the behavioral sciences* 1:24–38.

HARNAD, S.; DOTY, R. W.; GOLDSTEIN, L.; JAYNES, J.; and KRAUTHAMER, G., eds. 1977. *Lateralization in the nervous system.* New York: Academic Press.

HARRIS, MARVIN 1968. *The rise of anthropological theory: a history of theories of culture.* New York: Thomas Y. Crowell.

HARTLEY, DAVID 1810. *Observations on man, his frame, his duty, and his expectations.* 5th ed. 5 vols. Bath and London: n.p. 1st ed., 1749.

HASTIE, WILLIAM, ed. and trans. 1900. *Kant's cosmogony, as in his* Essay on the retardation of the rotation of the earth *and his* Natural history of the heavens. Glasgow: James Maclehose and Sons.

HENDEL, CHARLES WILLIAM 1934. *Jean-Jacques Rousseau: moralist.* 2 vols. London and New York: Oxford U. Press.

HERDER, JOHANN GOTTFRIED VON n.d. *Outlines of a philosophy of the history of man.* Trans. T. Churchill. New York: Bergman. Facsimile of 1800 London ed. 1st German ed., 1784.

1966. Essay on the origin of language. Trans. Alexander Gode. In *On the origin of language: two essays,* J. J. Rousseau and J. F. Herder, pp. 87–166. New York: Frederick Ungar. Written late 1770, publ. 1772.

HEWES, GORDON 1977. Language origin theories. In *Language learning by a chimpanzee,* ed. D. Rumbaugh, pp. 3–53. New York and London: Academic Press.

HILGARD, ERNEST R. 1960. Psychology after Darwin. In *Evolution after Darwin,* ed. Sol Tax, vol. 2, pp. 269–287. Chicago: U. of Chicago Press.

HJELMSLEV, LOUIS 1950–1951. Commentaires sur la vie et l'oeuvre de Rasmus Rask. *Conférences de l'Institut de Linguistique de l'Université de Paris* 10:143–157. Reprinted in *Portraits of linguists,* ed. T. A. Sebeok. Bloomington: Indiana U. Press, 1966.

HOARE, MICHAEL E. 1976. *The tactless philosopher: Johann Reinhold Forster, 1729–1798.* Melbourne: Hawthorne Press.

HODGEN, MARGARET TRABUE 1964. *Early anthropology in the sixteenth and seventeenth centuries.* Philadelphia.

HOFSTADTER, RICHARD 1945. *Social Darwinism in American thought, 1860–1915.* Philadelphia: U. of Pennsylvania Press.

HOLLOWAY, RALPH L. 1966. Cranial capacity, neural reorganization, and hominid evolution: a search for suitable parameters. *American Anthropologist* 68:103–121.

HUNT, JAMES 1865. On the Negro's place in nature. *Memoirs of the Anthropological Society of London* 1:1–64.

HUMBOLDT, WILHELM VON 1971a. *Diversity of the structure of human language.* Trans. F. A. Raven, Coral Gables, Fla.: U. of Miami Press.

———. 1971b. *Linguistic variability and intellectual development.* Trans. George C. Buck and Frithjor Raven. Coral Gables, Fla.: U. of Miami Press. 1st ed., 1836.

HUXLEY, THOMAS HENRY 1959. *Man's place in nature.* Ann Arbor: U. of Michigan Press. 1st ed., 1863.

JACOB, FRANÇOIS 1977. Evolution and tinkering. *Science* 196:1161–1166.

JAMES, WILLIAM 1890. *The principles of psychology.* 2 vols. London: Macmillan.

JEFFERSON, THOMAS 1955. *Notes on the state of Virginia.* Chapel Hill: U. of North Carolina Press. 1st ed., 1787.

JENKINS, P. F. 1977. Cultural transmission of song patterns and dialect development in a free-living bird population. *Animal Behavior* 25:50–78.

JERISON, HARRY J. 1973. *Evolution of the brain and intelligence.* New York and London: Academic Press.

JORDON, WINTHROP D. 1969. *White over black: American attitudes toward the Negro, 1550–1812.* Baltimore, Md.: Penguin Books. 1st ed., 1968.

KANT, IMMANUEL 1950. On the different races of men. In *This is race,* ed. Earl W. Count, pp. 16–24. New York: Henry Schuman. 1st German ed., 1775.

KEEN, BENJAMIN 1971. *The Aztec image in Western thought.* New Brunswick, N.J.: Rutgers U. Press.

KEITH, A. 1948. *A new theory of human evolution.* London: Watts.

KERR, ROBERT, ed. 1811. *Memoirs of the life, writings, and correspondence of William Smellie.* 2 vols. Edinburgh: John Anderson. Reprinted 1974, New York and London: Garland.

KROEBER, A. L. 1917. The superorganic. *American Anthropologist* 19:163–213.

KROODSMA, DONALD E. 1977. A re-evaluation of song development in the song sparrow. *Animal Behaviour* 25:390–399.

KUČERA, HENRY, and FRANCIS, W. NELSON 1967. *Computational analysis of present-day American English.* Providence, R.I.: Brown U. Press.

KUPER, ADAM 1973. *Anthropologists and anthropology: the British school, 1922–1972.* New York: Pica Press.

LAMARCK, J. B. 1914. *Zoological philosophy.* Trans. E. Eliot. London: Reprinted 1968, Hafner.

LEICESTER, HENRY M. 1974. *Development of biochemical concepts from ancient to modern times.* Cambridge, Mass.: Harvard U. Press.

LE GROS CLARK, W. E. 1963. *The antecedents of man.* New York: Harper and Row.

LEMAY, MARJORIE 1976. Morphological cerebral assymmetries of modern man, fossil man, and nonhuman primate. *Annals N.Y. Acad. Sciences* 280:349–366.

LENNEBERG, ERIC 1967. *The biological foundations of language.* New York: Wiley.

LÉVI-STRAUSS, CLAUDE 1962. Jean-Jacques Rousseau, fondateur des sciences de l'homme. In *Jean-Jacques Rousseau, fondateur des sciences de l'homme,* ed. Samuel Band-Bovy, pp. 239–248. Neuchatel: La Baconniere.

LIBERMAN, A. M. 1970. The grammars of speech and language. *Cognitive Psychology* 1:301–323.

1974. The specialization of the language hemisphere. In *The Neurosciences: A third study program,* ed. Schmitt and Worden, pp. 43–56. Cambridge, Mass.: MIT Press.

LIEDERMAN, PHILIP 1975. *On the origins of language.* London: Macmillan.

LOCKE, JOHN 1975. *An essay concerning human understanding.* Ed. Peter H. Nidditch. Oxford: Clarendon Press.

LORENZ, KONRAD 1957. The nature of instinct. Trans. Claire H. Schiller. In *Instinct: the development of a modern concept,* ed. C. H. Schiller, pp. 129–175. London: Methuen. 1st published in German, 1937.

LOVEJOY, ARTHUR O. 1923. The supposed primitivism of Rousseau's *Discourse on Inequality. Modern Philology* 21:165–186.

1933. Monboddo and Rousseau. *Modern Philology* 30:275–296.

1936. *The great chain of being.* Cambridge: Harvard U. Press.

LOVEJOY, A. O., and BOAS, G. 1935. *Primitivism and related ideas in antiquity.* Baltimore: Johns Hopkins Press.

LOWIE, ROBERT 1914. Social organization. *American Journal of Sociology* 20:68–97. Reprinted in *Lowie's selected papers in anthropology,* ed. C. DuBois, pp. 17–47. Berkeley and Los Angeles: U. of California Press, 1960.

1937. *The history of ethnological theory.* New York: Farrar and Rinehart.

LUBBOCK, SIR JOHN 1869. *Pre-historic times, as illustrated by ancient remains and the manners and customs of modern savages.* 2nd ed. London: Williams and Norgate. 1st ed., 1865.

LUKES, STEVEN 1972. *Émile Durkheim: his life and work.* New York: Harper and Row.

LYELL, SIR CHARLES 1873. *The geological evidences of the antiquity of man, with an outline of glacial and post-Tertiary geology and remarks on the origin of species.* 4th ed., rev. London: John Murray. 1st ed., 1863.

McDOUGALL, WILLIAM 1920. *An introduction to social psychology.* 15th ed. London: Methuen.

McGEE, W. J. 1901. Man's place in nature. *American Anthropologist* 3, no. 1:1–13. Reprinted in *Readings in the history of Anthropology,* ed. Regna Darnell, pp. 235–244. New York: Harper and Row, 1974.

McLENNAN, JOHN FERGUSON 1876. *Studies in ancient history, comprising a reprint of primitive marriage.* London: Bernard Quaritch.

MacLEAN, PAUL D. 1949. Psychosomatic disease and the 'visceral brain': recent developments bearing on the Papez theory of emotion. *Psychosomatic Medicine* 11:338–353. Reprinted in *Basic readings in neuro-psychology,* ed. Robert L. Isaacson. N.Y.: Harper and Row, 1964.

MAGOUN, H. W. 1960. Evolutionary concepts of brain function following Darwin and Spencer. In *Evolution after Darwin,* ed. Sol Tax, vol. 2, pp. 187–209. Chicago: U. of Chicago Press.

MAINE, SIR HENRY J. S. 1897. *Ancient law: its connection with the early history of society and its relation to modern ideas.* 6th ed. London: Murray.

MALINOWSKI, BRONISLAW 1960. *A scientific theory of culture and other essays.* New York: Oxford U. Press. 1st ed., 1944.

MALONE, KEMP 1952. Rasmus Rask. *Word Study* 28:1–4. Reprinted in *Portraits of linguists: a biographical source book for the history of Western linguistics, 1746–1963,* vol. 1, ed. T. A. Sebeok: pp. 195–199. Bloomington: Indiana U. Press, 1966.

MALTHUS, THOMAS R. n.d. *An essay on population.* 2 vols. London: J. J. Dent. Reprint of 7th ed. 1st ed., 1798.

MANUEL, FRANK E. 1959. *The eighteenth century confronts the gods.* Cambridge: Harvard U. Press.

⸻ 1965. *Shapes of philosophical history.* Stanford: Stanford U. Press.

MARLER, PETER 1970. Birdsong and speech development: could there be parallels? *American Scientist* 58:669–673.

⸻ 1975. On the origin of speech from animal sounds. In *The role of speech in language,* ed. J. F. Kavanagh and J. E. Cutting, pp. 11–37. Cambridge: MIT Press.

MARLER, PETER, and MUNDINGER, PAUL 1971. Vocal learning in birds. In *Ontogeny of vertebrate behavior,* ed. Howard Moltz, pp. 389–450. New York: Academic Press.

MARLER, PETER, and PETERS, SUSAN 1977. Selective vocal learning in a sparrow. *Science* 198:519–521.

MARX, KARL 1974. *The ethnological notebooks (1880–1882): studies of Morgan, Maine, Phear and Lubbock.* L. Krader, ed. Assen, Neth.: Van Gorcum.

MASTERTON, R. B.; CAMPBELL, C. B. G.; BITTERMAN, M. E.; and HOTTON, NICHOLAS, eds. 1976. *Evolution of brain and behavior in vertebrates.* Hillsdale, N.J.: Erlbaum; New York: Halsted.

MASTERTON, R. B.; HODOS, WILLIAM; and JERISON, HARRY, eds. 1976. *Evolution, brain, and behavior: persistent problems*. Hillsdale, N.J.: Erlbaum; New York: Halsted.

MATTINGLY, IGNATIUS G. 1972. Speech cues and sign stimuli. *American Scientist* 60:327–337.

MAUPERTIUS, P. L. MOREAU DE 1974. *Oeuvres*. 4 vols. New York and Hildesheim, Ger.: Georg Olms Verlag.

MAYR, ERNST 1963. *Animal species and evolution*. Cambridge: Harvard U. Press.

1972. The nature of the Darwinian revolution. *Science* 176:981–989.

MAZLISH, BRUCE 1975. *James and John Stuart Mill: father and son in the nineteenth century*. New York: Basic Books.

MEEK, RONALD L., ed. and trans. 1973. *Turgot on progress, sociology, and economics: a philosophical review of the successive advances of the human mind; on universal history, and reflections on the formation and the distribution of wealth*. Cambridge: Cambridge U. Press.

1976. *Social science and the ignoble savage*. Cambridge: Cambridge U. Press.

MEYNERT, THEODOR 1885. *Psychiatry —a clinical treatise on diseases of the forebrain, based upon a study of its structure, functions, and nutrition*. Part I: *The anatomy, physiology, and chemistry of the brain*. Trans. B. Sachs. New York: Putnam. Facsimile ed., New York and London, Hafner, 1968.

MILL, JAMES 1869. *Analysis of the phenomena of the human mind*. 2 vols. London: Longmans Green Reader and Dyer. 1st ed., 1829.

MILL, JOHN STUART 1965. *Principles of political economy, with some of their applications to social philosophy*. Toronto: U. of Toronto Press; London: Routledge and Kegan Paul.

1969. *Utilitarianism. Essays on ethics, religion, and society. Collected works of J. S. Mill*. Vol. 10. Pp. 204–259. Toronto: U. of Toronto Press; London: Routledge and Kegan Paul. First published, 1861.

MITCHELL, ARTHUR 1880. *The past in the present: what is civilization?* Edinburgh: David Douglas.

MONBODDO, LORD (JAMES BURNET) 1774–1792. *Of the origin and progress of language*. 2nd ed. 6 vols. Edinburgh: J. Balfour. Facsimile ed. of 2nd ed., 6 vols., New York, Garland, 1970.

MONTAIGNE, MICHEL DE 1967. *Essays*. Trans. J. M. Cohen. Harmondsworth, Eng.: Penguin. 1st French ed., 1580.

MONTESQUIEU, BARON DE 1900. *The spirit of laws*. Trans. Thomas Nugent. 2 vols. London and New York: Colonial Press.

1964. *Lettres Persanes*. Paris: Garnier-Flammarion. 1st ed., 1721.

MORGAN, C. LLOYD 1913. *Instinct and experience*. London: Methuen. 1st ed., 1912.

MORGAN, LEWIS HENRY 1964. *Ancient society.* Cambridge: Belknap Press. 1st ed., 1877.

MOSSE, G. L. 1963. *The culture of western Europe: the nineteenth and twentieth centuries.* London: John Murray.

MÜLLER, J. 1843. *Elements of physiology.* Trans. William Baly. Philadelphia: Lea and Blanchard. Arranged from the 2nd London edition by John Bell.

NOTTEBOHM, FERNANDO 1970. Ontogeny of Bird Song. *Science* 167:950–956

1972. The origins of vocal learning. *American Naturalist* 106:116–140.

1977. Asymmetries in neural control of vocalization in the canary. In *Lateralization in the nervous system,* ed. Steven Harnad et al., pp. 23–44. New York and London: Academic Press.

PARKS, G. B. 1964. A turn to the romantic in the travel literature of the eighteenth century. *Modern Languages Quarterly* 25:22–33.

PASSINGHAM, R. E. 1973. Anatomical differences between the neocortex of man and other primates. *Brain, Behavior and Evolution* 7:337–359.

PEDERSEN, HOLGER 1931. *The development of linguistic science in the nineteenth century.* Trans. J. W. Spargo. Cambridge, Mass.: Harvard U. Press. Reprinted Bloomington: Indiana U. Press, 1962.

PENNIMAN, T. K. 1935. *A hundred years of anthropology.* London: Duckworth.

POPE, ALEXANDER 1734. *An essay on man.* London: John Wright, Printer for Lawton Gilliver. Facsimile ed., Menston, Eng.: Scolar Press, 1969.

POPPER, KARL R. 1963. *Conjectures and refutations: the growth of scientific knowledge.* London: Routledge and Kegan Paul.

RADCLIFFE-BROWN, A. R. 1952. *Structure and function in primitive society.* London: Cohen and West.

RASK, RASMUS K. 1976. *A grammar of the Icelandic or old Norse tongue.* Trans. Sir George Webbe Dasent. Introduction by T. L. Markey. *Amsterdam Classics in Linguistics* vol. 2. Amsterdam: John Benjamins. Reprint of 1843 English trans. of 1818 Swedish ed.

RAUSCHENBERG, R. 1968. Daniel Carl Solander, naturalist on the *Endeavour. Trans. Amer. Phil. Soc.* 58 no. 8.

REYNOLDS, VERNON 1967. *The apes.* New York: E. P. Dutton.

RIVERS, W. H. R. 1924. *Instinct and the unconscious: a contribution to a biological theory of the psycho-neuroses.* Cambridge: Cambridge U. Press. 1st ed., 1920.

1926. *Psychology and ethnology.* London: Kegan Paul, Trench, Trubner.

ROBINSON, BRYAN W. 1976. Limbic influences on human speech. *Annals N.Y. Acad. Sciences* 280:761–771.

ROCHER, ROSANE 1968. *Alexander Hamilton (1762–1824): a chapter in the early history of Sanskrit philology.* New Haven, Conn.: American Oriental Society.

ROGER, JACQUES   1971. *Les sciences de la vie dans la penseé francaise du XVIIIe siècle: La generation des animaux de Descartes a L'Encyclopedie.* n.p.: Armand Colin.

ROUSSEAU, JEAN JACQUES   1817. *Oeuvres completes.* Paris: A. Belin.

1916. *The social contract and discourses.* Trans. G. D. H. Cole. London: J. M. Dent; New York: Everyman's Library, E. P. Dutton. Reprint of 1913 ed.

1964. *The First and Second Discourses.* Roger D. Masters, ed. Trans. Roger D. Masters and Judith R. Masters. New York: St. Martin's Press.

1975. *The confessions.* Trans. J. M. Cohen. Hammondsworth, Eng.: Penguin Books. 1st French ed., 1781.

SAPIR, EDWARD   1907–1908. Herder's 'Ursprung der Sprache.' *Modern Philology* 5, July:109–142.

SCHLEIDT, WALTER   1950. The concept of race in anthropology and the divisions into human races from Linnaeus to Deniker. In *This is race,* ed. Earl W. Count, pp. 354–391. New York: Henry Schuman.

SCHLEGEL, FRIEDRICH VON   1852. *The philosophy of history in a course of lectures delivered at Vienna.* 6th ed. Trans. James B. Robertson. London: Bohn.

SCHULTZ, A. H.   1976. The rise of primatology in the twentieth century. *Folia Primatol.* 26:1–17.

SILVERSTEIN, MICHAEL, ed.   1971. *Whitney on language: selected writings of William Dwight Whitney.* Cambridge, Mass.: MIT Press.

SIMPSON, G. G.   1953. *Major features of evolution.* New York: Columbia U. Press.

SIQUELAND, EINAR R., and DeLUCIA, CLEMENT A.   1969. Visual reinforcement of nonnutritive sucking in human infants. *Science* 165:1144–1146.

SMITH, ADAM   1812. *Works, with an account of his life and writings by Dugald Stewart.* 5 vols. London.

1967. Letter to the *Edinburgh Review.* In *The early writings of Adam Smith,* ed. J. Ralph Lindgren, pp. 15–28. New York: Augustus M. Kelley. First published, 1755.

1971. *The theory of moral sentiments.* New York: Garland. Facsimile ed. of 1759 London ed.

SMITH, BERNARD   1960. *European vision and the South Pacific 1768–1850: a study in the history of art and ideas.* Oxford: Clarendon Press.

SMITH, G. ELLIOTT   1924. *Essays on the evolution of man.* Oxford: Oxford U. Press.

SPENCER, HERBERT   1852. The development hypothesis. *The Leader,* March 20. Reprinted in *Essays: scientific, political and speculative,* vol. 1, pp. 1–7. London: Williams and Norgate, 1891.

1857. Progress: its law and cause. *Westminster Review,* April. Reprinted in *Essays: Scientific, political and speculative,* vol. 1, pp. 8–62. London: Williams and Norgate, 1891.

1870. *Principles of psychology*. 2nd ed. 2 vols. London: Williams and Norgate.

1885. *The principles of sociology*. 3rd ed. rev. Vol. 1; Vol. 6 of *A system of synthetic philosophy*. London: Williams and Norgate.

1892. *Social statics*. Rev. ed. London: Williams and Norgate. 1st ed., 1850.

STEVENS, KENNETH S. 1975. Speech perception. In *The nervous system*, vol. 3, ed. D. B. Tower, pp. 163–171. New York: Raven Press.

STOCKING, GEORGE W., Jr. 1968. *Race, culture, and evolution: essays in the history of anthropology*. New York: Free Press; London: Collier-Macmillan.

1971. What's in a name? The origins of the Royal Anthropological Institute (1837–71). *Man* n.s. 6:369–90.

TEGGART, FREDERICK J. 1960. *Theory and process of history*. Berkeley and Los Angeles: U. of California Press. Reprint of *Theory of history*, New Haven, Conn.: Yale U. Press, 1925, and *Processes of history*, New Haven, Conn: Yale U. Press, 1918.

TINBERGEN, N., and PERDECK, A. C. 1950. On the stimulus situation releasing the begging response in the newly-hatched herring gull chick *(Larus argentatus)*. *Behaviour* 3:1–38.

TOULMIN, STEPHEN, and GOODFIELD, JANE 1965. *The discovery of time*. London: Hutchinson.

TYLOR, EDWARD B. 1889. *Anthropology: an introduction to the study of man and civilization*. 2nd ed. rev. London: Macmillan.

TYSON, EDWARD 1699. *Orang-outang, sive homo sylvestris: or, the anatomy of a pygmie compared with that of a monkey, an ape, and a man*. London: T. Bennett and D. Brown. Facsimile ed., Folkstone, Kent: Dawson, 1966.

VARTANIAN, ARAM, ed. 1960. *La Mettrie's l'homme machine: a study in the origins of an idea*. Princeton: Princeton U. Press. 1751 ed. included, in French.

VAUGHN, C. E., ed. 1962. *The political writings of Jean Jacques Rousseau*. 2nd ed. 2 vols. In French. Oxford: Blackwell.

VERBUG, P. A. 1950. The background to the linguistic conceptions of Franz Bopp. *Lingua* 2:438–468. Reprinted in T. A. Sebeok, ed., *Portraits of linguists: a biographical source book for the history of Western linguistics, 1746–1963*, vol. 1, ed. T. A. Sebeok, pp. 221–250. Bloomington: Indiana U. Press.

WATSON, JOHN B. 1914. *Behavior: an introduction to comparative psychology*. New York: Henry Holt.

1970. *Behaviorism*. New York: Norton. 1st ed., 1924.

WESTERMARCK, EDWARD 1921. *The history of human marriage*. 5th ed. 3 vols. London: Macmillan.

1926. *A short history of marriage*. London: Macmillan.

WHITNEY, WILLIAM DWIGHT 1883. *Life and growth of language*. 4th ed. London: Kegan Paul, Trench. 1st ed., 1875.

1971. *Language and the study of language.* New York: AMS Press. Reprint of 6th ed., 1901. 1st ed., 1867.

WILKIE, J. S. 1959. Buffon, Lamarck, and Darwin: the originality of Darwin's theory of evolution. In *Darwin's biological work, some aspects reconsidered,* ed. P. R. Bell et al. pp. 262–307. Cambridge: Cambridge U. Press.

WILSON, ARTHUR M. 1957. *Diderot. The testing years, 1713–1759.* New York: Oxford U. Press.

WOODWORTH, R. S. 1952. *Contemporary schools of psychology.* 8th ed. London: Methuen.

YANKELOVITCH, D., and BARRETT, W. 1970. *Ego and instinct.* New York: Random House.

YATES, FRANCES A. 1964. *Giordano Bruno and the Hermetic tradition.* Chicago: U. of Chicago Press; London: Routledge and Kegan Paul.

YOUNG, ROBERT M. 1970. *Mind, brain and adaptation in the nineteenth century: cerebral localization and its biological context from Gall to Ferrier.* Oxford: Clarendon Press.

2

MORPHOLOGY AND
FUNCTION

DISCOURSE ON METHOD

If anthropological theory has implicitly accepted a stratified conception of man and nature, and if the development of postwar psychobiology has called these assumptions into question, then an attempt to refute the claims of biological determinants of human behavior by an appeal to cultural levels of organization and essential differences between man and animal must necessarily fail, because such categorical distinctions presuppose exactly the kind of a priori divisions that cannot be accommodated to the already existing data on apes. The question is fundamentally a question of assumptions and not a dispute about facts. Even people who have devoted their lives to the study of apes have been led by their assumptions to theoretical positions that fail to take adequate account of the known facts of primate psychology. A professor of anthropology at Oxford, for example, whose research into the history of primatology was cited in the previous chapter, has written within the past few years that human and animal societies are essentially different because of the absence of animal conceptualization (V. Reynolds 1976, pp. 29–30):

> The first point about animal society is an absolutely crucial one. It is that animals (by this I mean all species other than man) do not think conceptually; they respond to stimuli or configurations of stimuli coming within their own bodies and from the surrounding environment, but do not conceptualize either themselves, or others, or

the external world and its parts, or their social group. The reason is that animals other than man lack symbolic language-systems needed to produce the cognitive constructions that are necessary for symbolic concept formation.

It is characteristic of summations derived from underlying assumptions that they accept as established what is precisely most problematical, and the investigation of primate conceptualization, although far from advanced, has had its noteworthy successes, discussed in chapter 4, which vitiate such claims of absolute differences. Nonetheless, an interactionist conception of psychobiology and human behavior is a two-way street, and a good case can be made that classical ethology was founded on exactly the same premise as social anthropology: an absolute distinction between innate and acquired.

Nineteenth century progressivist psychology had maintained that instinct evolves into reason phylogenetically and that reason replaces instinct. Konrad Lorenz, in developing the descriptive biology of instinctive behavior, was at pains to show that it formed a distinct category of phenomena that did not "merge imperceptibly" into reflex on one hand and learning on the other. In stressing this distinctiveness of instinctive behavior, he placed great theoretical weight on stereotyped behavior patterns that appear endogenously motivated and fully formed in animals reared in social isolation. These *fixed-action patterns*, as they are called, are so invariant that they can be used as taxonomic features for classifying the species that exhibit them (1958, 1973; McKinney 1964). Lorenz credits Otto Heinroth and Charles Whitman, both ornithologists working in the early decades of the twentieth century, with the discovery that instinctive acts were as amenable to classification by comparative methods as any feature of anatomy. Whitman produced a taxonomy of pigeons that put behavior and morphology on an equal footing, and Heinroth, with a flash of historical perspicacity rivaling Freud, pointed out that most birds scratch their heads like four-legged animals: they extended the wing, which corresponds to the forelimb, before moving the foot (the hindlimb) to the head, even though they could easily reach the head by moving the leg medially and keeping the wing where it was. From the perspective of intellectual history, it does not matter whether or not this is a true generalization about birds; it is more important that the persistence of this pattern was recognized as a historical rather than a functional phenomenon. Without using the term, both Heinroth and Whitman realized the possibility of establishing homologies in behavior that were fully comparable to anatomical homologies. Lorenz extended Heinroth's observations on ducks to systematic comparisons of courtship displays, and he was able to show that the displays of related species were built out of similar innate behavior patterns that had been quantitatively and qualitatively altered. In this paper, which begins with an oblique reference to historical linguistics, Lorenz demonstrated that homologous behaviors in different species are not necessarily identical but rather *systematically different*.

In the light of linguistic history, the reference to historical linguistics is not fortuitous, for Lorenz was applying exactly the same method to animal behavior that had been applied to human phonology a century before. Yet, as powerful as the "argument of homology" can be as an analytic device, it rests on an assumption of morphological invariance in the behaviors being compared and on a very direct coupling between the form of the behavior and its underlying genetic control. Since classical ethology made the fixed-action pattern into one of the defining features of instinctive behavior, it had the paradoxical result of reasserting some of the major premises of nineteenth century thought in the context of modern biology. Although Lorenz explicitly stated that instincts have a developmental history, which allows for ontogenetic modification, and even though he formulated the concept of instinct-learning interlocking, discussed in chapter 1, the homological method requires that these modifications be given ancillary status in the theory of instinct. The theoretical difficulties of classical ethology were compounded by another defining property of instinctive behavior: its endogenous motivation.

In formulating the theory of instinct, classical ethology drew upon the nineteenth century concept of nerve energy or nerve force, itself derived from the hydraulic model of electrical energy as formulated by physics (Darwin 1872; Muller 1843).[1] In Konrad Lorenz's hands, there were as many forms of "nerve energy" as there were endogenously activated instincts: each instinct was motivated by what he termed its own *action-specific energy*. Lorenz explained variation in instinctive acts by differences in the quantity of energy. At low levels of action-specific energy, the instinctive sequence may appear in an incomplete form. Thus, portions of the nesting behavior of the graylag goose can appear all year long but they only become exhibited as part of a full-fledged instrumental sequence during the nesting season. Quantitative variations in action-specific energy could lead to very different behavioral manifestations, because instinctive acts often include dissimilar behaviors as part of a series, but at high levels of energy the dissimilarities will be revealed as constituents of a superordinate instinctive act. Although at low energy levels, the behavioral components may appear as purposeless and nonadaptive, they will reveal their adaptiveness if examined in the appropriate context. However, in Lorenz's formulation, instincts were not simply innately motivated innate behavior patterns, but the motivational component interacted with innate perceptual components, called releasing stimuli, which normally triggered them off. Ethologists have demonstrated many examples of releasing stimuli in the behavior of animals, and Lorenz showed that behavior patterns could be released *in vacuo* if deprived of the releasing stimuli for a sufficient length of time. Lorenz's concept of action-specific energy and the consummatory behavior that satisfied a particular innate motivation led him to an important criticism of traditional instinct theory. As Lorenz emphasized, the goal of an instinctive act is not the effects it produces but the discharge of the behavior itself. A squirrel is not trying to bury nuts but releasing the tension created by the accumulated instinctive motivation. European squirrels (*Sciureus vulgaris*) will in fact try to "bury nuts" even if raised in social isolation on liquid food on a cement floor (Eibl-Eibesfeldt 1975). The instinct theory of the late nineteenth century, which categorized instincts in terms of biological functions, such as reproduction of the species, confused the biologist's theoretical interpretations of the function of the act with the subjective psychology of the animal. The instinctive act is independent of its biological value, and one consequence is that satisfaction of the biological function through other methods has no effect on the probability of the instinctive acts that ordinarily subserve this function. Feeding a cat with canned cat food does not diminish its hunting behavior. Lorenz criticizes McDougall's list of human instincts on precisely the ground that biological functions are being confused with psychological processes (McDougall 1908).

If such an instinct theory is applied to primates, however, then one is forced to postulate two distinct spheres of behavior: an instinctive domain, in which the external consequences of an action are irrelevant to its performance, and an intelligent domain based on the analysis of the external situation and the choice of relevant strategies. Such a division, which parallels the classical ethological distinction between appetitive and consummatory behaviors, has in practice little to distinguish it from nineteenth century progressivism. Lorenz, in stressing that instinct was a distinct category that did not grade phylogenetically into learning, concluded that "in higher vertebrates the evolution of intelligent behavior has definitely gone hand in hand with a corresponding devolution of instinctive activities." (1957, p. 151). Also in Eibl-Eibesfeldt's massive compendium, which defines ethology as "the biology of behavior" and sets out to demonstrate the relevance of the theory to human action, we come away with the conclusion that higher primates are particularly well characterized by the faculty of insight and their innate capacity to learn — a conclusion indistinguishable from the major premise of the ladder of reason (1975). In a volume designed to bridge the conceptual gap between biology and culture, Konrad Lorenz takes an anthropological position and argues for a cultural level of ritualization, which is analogous to biological ritualization down to the last detail (Huxley, 1966).[2] Both cultural and biological rituals are communicative; they are modified from their prototypical movements in the speed, amplitude, substitution, reduction, and repetition of components; they incorporate previously distinct elements into a new context; they develop a motivational impetus of their own; and they perform the same functions of social bonding and canalizing aggression. Lorenz notes by way of explaining this remarkable convergence that the nervous system has "an insidious way of performing analogous functions on different levels of integration in such similar manner as to mislead even the sophisticated observer to assume physiological identity where it does not exist" (1966, p. 279). Yet in viewing these similarities, one suspects that the insidious process at work is the act of cutting a single phenomenon into two equal parts and then marveling at the symmetry of the halves.

Lorenz has shifted his frame of reference between the initial premise and the final conclusion. Because learned behavior does not derive from instinctive behavior phylogenetically — which is to say that they have distinct neural substrates that are not homologous — it does not follow that rituals that have learned components must therefore be mediated by the noninstinctive parts of the brain. There is another alternative: that the distinction between innate and learned behavior, which seems so apparent in lower vertebrates, is replaced in primates by a capacity for the developmental interaction of innate and environmental components. The distinction between innate and learned does not become meaningless, as some behavior patterns can develop in social isolation, nor does learning replace instinct. However, a division into learned and innate can only be used to characterize the ontogenetic history of particular behaviors considered out of context, whereas the *functional* characterization of primate behavior requires a model of ontogenetic interaction between environmental information and innate components.

The question here is not whether relatively stereotyped and innate behaviors, corresponding to fixed-action patterns, occur in primates. There is evidence from both isolation rearing experiments and from neurophysiological research that fixed-action patterns occur in nonhuman primates and man. Primates reared in social isolation have a large motivational and behavioral repertory, to be discussed in chapter 3, and artificially induced brain lesions show that many of the motor components of behavior are organized by mechanisms in the caudal parts of the mammalian brain (Bard and Mountcastle 1948; Bard and Macht 1958; Von Holst and Von St. Paul 1963; Akert 1961; Hess 1957).[3] Cats that have had all the brain removed above the level of the midbrain, except for the endocrine tissue of the hypothalamus, which is left neurally disconnected, are still capable of various kinds of behavior. They can walk if stimulated, although they do not do it spontaneously, and they show behavioral components of fear and rage. A painful stimulus can produce hissing, growling, tail lashing, claw protrusion, turning, and walking. Similarly, fear behavior, with crawling locomotion, hair erection, and dilated pupils can also be provoked. Compared to the displays of animals with more anterior (rostral) portions of the brain remaining, the displays are incomplete. However, the cortex is not necessary for complete displays of aggression in cats or dogs, and attack and aggressive display can both be elicited from the midbrain in cats by electro-stimulation. The sexual behavior of courtship and copulation can also be induced by hormone treatments of cats that have all the brain removed above the level of the hypothalamus.

The finding that the fixed-action patterns of aggression, sex, and fear are organized in the midbrain or more caudal areas in cats and rats is applic-able to primates as well. Aggression can also be elicited by electrostimula-tion of the hypothalamus in monkeys, and lesions of the caudal parts of the brain in man can cause pathological laughter and crying, which occur as motor release phenomena not associated with the corresponding affective states. Moreover, humans born without a forebrain are still capable of a wide range of reflexively organized behaviors. These findings support the concept of innate fixed-action patterns in mammals, but to argue that they support the classical theory of instinct in primates is to take insufficient notice of the dramatic changes in the neural context of brainstem mechanisms.

As is generally acknowledged, the primate order has been charac-terized by an expansion of the brain generally and by an expansion of the neocortex in particular. The homologies of morphology and mechanism well established by neurophysiological study of the caudal parts of the vertebrate brain must be reconciled with these equally striking progressive changes. The nineteenth century solution was to view the nervous system as a hierarchy of neural control, with each level superimposed on the one below it. This Victorian brain, with the neocortex superimposed on a hierarchy of functional levels, each with properties of the reflex arc, has given way to a functional interaction among anatomically distinct compo-nents, in which phylogenetically different levels are integrated by complex loops of neurochemically modulated circular circuitry (Kornhuber 1974).[4] Also, the same caudal regions of the brain that integrate the part behaviors of affect also give rise to fiber tracts that project to wide areas of the forebrain and influence global brain states such as sleep and alertness through neu-rochemical modulation. Thus, there is a functional and bidirectional inter-action between phylogenetically primitive and advanced structures in the brain, and these facts are not easily reconciled with a theory of behavior that forces us to accept instinct as one thing and intelligence as another.

The limitations of the theory of classical ethology, when applied to primates, are due precisely to the factors that contributed to its success with lower vertebrates. If instinct is defined by its behavioral invariance and by its imperviousness to assessments of external efficacy, then instinct must necessarily contract as intelligence advances. As with the Ladder of Reason, this theory reduces instinct in man to a dismal litany of coughing, sneezing, and the last six seconds of mating behavior. It is no wonder that social scientists have not been eager to follow ethology down this particular path.

Even worse, the ethologist, armed with the fixed-action pattern as his only cross-cultural tool, has made a determined effort to show that they are everywhere the same. If a smile is a smile is a smile, then once again we are faced with a uniform human nature draped in the cloak of custom, and ethology, by its own premises, is irrelevant to the human variety that in fact exists.

In attempting to deal with the larger issues of human culture, classical ethology has failed to perceive that its attempts have contradicted the theoretical premises it purports to hold. For example, Eibl-Eibesfeldt and Wickler have pointed out the morphological similarities between the innate fixed-action patterns of nonhuman animals and the cultural products of man: the penile displays of adult male squirrel monkeys and the phallic emblems that adorned houses in classical civilization, and the crests of birds and the plumes of military hats (Wickler 1967; Eibl-Eibesfeldt 1975). These parallels do not satisfy any ethological definition of an instinctive act: they are culturally diverse, constructed by learned sequences of behavior, and often irrelevant to the opportunity to engage in the consummatory acts appropriate to the motivations they mimic. If these constructed symbols are examples of human instinctive behavior, then human instinct must be quite different from what ethological theory would lead us to expect. But conversely, if human beings are using their phylogenetically advanced ability to creatively transform matter to simulate the innate displays of animals, then there are clearly some peculiarities of human intelligence not adequately encompassed by the devolution of instinct.

A nineteenth century theory, with its "levels of organization" existing as closed Comtian spheres with a science appropriate to each, would have no trouble whatsoever in reconciling this juxtaposition of penile displays and classic statuettes. To the contrary, such parallelisms would be regarded as proof-positive for the cultural integration in man of mechanisms only biologically elaborated in his predecessors. To the nineteenth century theorist, a better example of the *convergence* of the cultural and biological levels of organization would be difficult to find. It is only in the context of postwar psychobiology that the sweeping explanatory power of cultural-biological convergence can be seen as problematical. If, as we have argued, this neat distinction of levels is the product of a static evolutionism transposed to man and nature, then the levels have not been demonstrated but only assumed. The distinction does not rest on validated differences between humans and other animals but on a hierarchical concept of the relationship between man and nature that owes nothing to the study of primates whatsoever. Consequently, an a priori cultural level of organization cannot be used to refute the applicability of biological factors in the determination of human institutions, as some anthropologists have done, but neither is anyone required to accept biological theories of human behavior deduced from prevailing notions of the gene in the absence of mechanisms for their implementation (Sahlins 1976; Wilson 1975). Behavioral primatology proceeds from the premise that the nature of human nature will be the last thing that human science will discover. As such, it cannot dismiss the parallelisms of penile displays and phallic emblems or the detailed similarities in process between innate and cultural rituals as mere convergences that validate the distinct domains of biology and culture.

If the theory of evolution is true, then the behavioral differences among primates must be explicable as the product of transformation over time. However, the anthropological strategy, reinforced by Comte's hierarchy of the sciences, has been to regard these transformations as leading to a level of organization in man that can be explained without recourse to these so-called lower levels: an evolutionary emergent of collective representations, social facts, conventions, and institutions. The existence of such phenomena is not in question, but the methodological claim that such phenomena constitute a level of organization that can be understood without reference to psychobiological variables is a very strong claim indeed, and it is not easily reconciled with the findings of natural science. If one seeks to understand the phenomenon of visual perception, for example, meaningful propositions can be articulated about many different levels of organization simultaneously: physicochemical interaction between a molecule of rhodopsin and a quantum of light, histological characterization of the differences among retinal receptor cells, electrochemical events at the synapse between nerve cells, neurophysiological characterization of patterns of inhibition and facilitation among neighboring cells, anatomical delimitation of the pathways of interconnection, psychophysical recording of the behavioral responses of the intact system to systematically varied input, gestalt psychological determinations of perceptual set, historical considerations about the presence or absence of a tradition of perspective art, and so on. The claim that complex functional systems, which embrace many levels of organization simultaneously, can be understood by regarding each level as causally self-contained is simply an a priori methodological assumption, and, like most assumptions, it must be evaluated by the degree of understanding it makes possible.

Because the methodology of social anthropology assumes the existence of the very levels of organization which any human evolutionary theory must explain, it is an untenable methodology when viewed from the perspective of evolutionary theory and psychobiology. At the very least, the development of an evolutionary theory of human behavior requires that the methodological claim of social anthropology be rejected as a guiding methodology for evolutionary primate studies. However, it would be a mistake to think that a rejection of the methodological claims of social anthropology is simply a gentleman's disagreement about the boundaries of basically complementary approaches to a complex reality. Since an evolutionary theory of human behavior has as its major goal the conceptualization of the hominid emergence, a school devoted to such a theory cannot simply accept at face value the characterizations of human institutions bequeathed by social anthropology. To accept Durkheim's definition of a social fact or a collective representation is to accept a definition of human institutions that forecloses any possible reconciliation between the behavior of humans and animals. Although there is no doubt that human beings often like to think of institutions as structures existing apart from themselves—as continuous as the apostolic succession and as immutable as the Nuer—such a conception of institutions is so transparently the ideology of ruling elites transposed to the phenomenal realm that primate psychobiology is under no compulsion to take it seriously. Evolutionary theory need only take seriously the fact that human beings find such reifications particularly compelling.

A primatological approach rejects the sui generis social fact and the disembodied collective representation; it begins instead with the heuristic that institutional behavior in man is the *product of the interaction* among individuals and groups seeking to implement their own motives, through instrumental action, in accommodation to the perceived behavior of others, as interpreted by conceptually encoded information as to nature of the ongoing interaction and their relation to it, and partially mediated through innate behavior patterns experientially modified. In such an approach, institutions are still the coordinated social action by which humans produce a collective product not reducible to the individual action of the participants; but it does not suppose that such collective action can exist in a realm causally independent of the knowledge, skills, feelings, and aspirations of the people who bring it about, or that a characterization of the collective product is a substitute for explicit treatment of the behavioral coordinations allowing such complex products to be created and sustained. In approaching the study of institutions from the bottom up, through the actual social interactions by which they are implemented, the characterization of institutions falls within the purview of those psychobiological determinants of behavior that affect primate interaction generally. Indeed, such a conclusion should engender no great surprise to those who already think of institutions as historical processes and not as timeless structures. Even a cursory knowledge of human history suggests that collective enterprises no longer able to engage the positive feelings of their participants or to undertake the complex coordinations necessary for their existence fall into decay, and this in itself suggests that such purely ethological variables as affective display and innate motivations are not mere spectators to the rise and fall of human institutions but determinants to be reckoned with.

The institution conceived as interaction alters the strategy of evolutionary anthropology. We do not ask how a level of collective representation emerges in the course of evolution, but we do ask another question, superficially similar, which leads to quite different results: what are the behavioral and psychological differences between man and other primates, as revealed by comparative study, and what transformations are necessary in a human precursor to evolve the social coordination commensurate with the collective products of *Homo sapiens?* By this line of reasoning, the emergence of *Homo sapiens* does not require a theory of levels at all. It requires a theory of the transformation of psychobiological mechanisms, similar to those of nonhuman primates, so they become capable of mediating the kinds of social coordination we see in modern man. Institutions, in other words, do not appear with modern man, for all primates have coordinated social action. What is new in humans are some of the mechanisms of social coordination. These innovations in turn are based on evolutionary changes in the primate biogram; and social interactions among organisms so constituted give rise to products that are impossible to animals lacking these psychobiological innovations. However, this perspective is far different from the claim that human social life is constituted on different principles from that of animals. It would be more correct to say that it is constituted on *additional* principles, but these additions can by no means be assumed to invalidate all that has gone before or to exist in biological vacuum.

THE LIMITS OF MORPHOLOGY

This tendency to see institutions as things rather than as processes has prevented the realization that many of the building blocks of an evolutionary anthropology have already been gathered. Ethological investigation, for example, has established the existence of fixed-action patterns in man, demonstrated the homologies among affective communication in man and animals, shown a capacity for observational learning in great apes, established the fact (but not the extent) of conceptual thought in animals, recorded many of the morphological similarities in the behavior and social organization of man and other primates, shown progressive evolutionary changes in self-conceptualization and tool-using ability, developed procedures for language-training in apes, created animal models for the interaction of innate and acquired information, applied the technology of film and video to the fine-grained study of social coordination, and demonstrated the mutual interaction of both phylogenetically primitive and advanced behavior in the course of social interaction. This is an impressive achievement, and even in its present incomplete state, it can be integrated into an evolutionary theory of human behavior, provided that we reject the characterizations of human nature by social anthropologists on one hand and classical ethologists on the other.

Fixed-action patterns are a good case in point. As reviewed in chapter 3, the initial anthropological response to the discovery of innate, stereotyped behaviors was a denial that they occurred, while classical ethology articulated its discovery in the context of such constructs as endogenous motivation and vacuum execution. Both of these positions make it difficult to accept the fact of fixed-action patterns without espousing a particular theoretical evaluation of them. A fixed-action pattern, like *parry,* which is the rapid movement of the arm to the front of the face to shield it from a blow, appears in both humans and chimpanzees, and it probably develops without specific learning, as do many other skeletal muscle behaviors, some much more complex, such as walking. Yet a theory that insists upon the innateness of fixed-action patterns as the primary fact would hardly predict a commonplace occurrence of parry as a gesture to mimic fear in a conversation between two people. From the perspective of primate interaction, it is quite easily demonstrable that fixed-action patterns, although still innate, are available for integration into complex behavioral structures controlled by phylogenetically progressive mechanisms. Yet the appositeness of the gesture is partially determined by the facts of its innate occurrence. It is a suitable gesture because a person can recognize its innate links to fear while *simultaneously* realizing, on the basis of other cues, that the sender is not afraid. Human social interaction uses such juxtapositions all the time, and as such, humans, and perhaps other higher primates, have carried Marler's principle of multimodal communication one step further: they not only convey an affect in many sensory modalities at once but they combine distinctly different levels of neural control into composite signals.

If we accept this proposition as a factual account of one way in which humans interact, then it is equally fatal to any theory that rests its case on the ultimate reality of morphological distinctiveness. A smile is not a smile is not a smile. Rather, humans are (1) determining the suitability of signals by (*a*) the integration of conceptually stored information about innate contexts with (*b*) conceptually mediated understanding of the ongoing interaction, and (2) producing a signal that executes (*a*) an innate motor program under (*b*) the control of volitional action. In this conceptualization, innate motor components and innate contexts of activation, such as releasing signals, are constantly being integrated with conceptually stored information to produce a composite product, and it is precisely this integration that gives the behavior of higher primates its subtlety and richness. In this characterization, which we assert as a second methodological heuristic, the properties of innate action, volitional control, and conceptual thinking are partial determinants of the behavioral product, and as such, both ethological and psychological theory are directly pertinent to our understanding of the processes of social coordination. Indeed, so little understanding has been derived from a century of social anthropology because the methodologies are so constituted as to deny that such interaction either occurs or is relevant. The insistence that psychobiology "be reconciled" with the findings of social anthropology is without force. It is rather the task of social anthropologists to abandon their reified institutions, divorced from all processes of psychobiological integration, and begin documenting the ways in which culturally variable behavior is coordinated in specific interactive contexts.

GETTING NETWORKS TOGETHER

The psychobiological investigations of primate behavior presently constitute a body of findings rather than an integrated theory, but these building blocks are not as chaotically strewn about as may at first appear. Some possibilities for theoretical integration are already discernible. For example, as indicated by studies such as Van Hooff's (1973; and chap. 3), it can be demonstrated that fixed-action patterns are not strictly segregated units but are organized into networks in which the activation of one pattern leads to the occurrence of other units to which it is related. The patterns relating to aggression, for example, tend to cluster together in time, as would be expected, and to preclude incompatible clusters like affiliation. These preferential associations among fixed-action patterns can be termed the *innate motor network,* and further support for it can be adduced by the studies of caudal brain mechanisms already mentioned. Electrostimulation studies in chickens, for example, show overlapped zones where distinct behavior patterns can be elicited. It is also known from ethological investigations that lower vertebrates have a repertory of innate perceptual units, known as innate releasing mechanisms, which are systematically interconnected with the motor network. The eye spot configuration discussed in chapter 3 or the sexual signals noted in chapter 7 are well-known examples of such innately salient perceptual cues. In the classical ethological interpretation, however, there is a very close relationship between the innate releasing mechanism on one hand and certain fixed-action patterns on the other: a particular signal elicits a particular response. Once again, however, invariance is being made a defining feature of instinct, and such close coupling between stimulus and response has very little explanatory power when applied to primates. This fact has been used to argue for the irrelevance of innate cues in primate behavior, but another interpretation is possible: that primates have an innate *perceptual network* as well as innate motor network, but that in primates the direct links have come to be nested within higher systems. From this perspective, the innate components of behavior and perception are alive and well, even in man, but these networks have become integrated with phylogenetically advanced mechanisms specialized for (1) the *contextualized activation* of innate connections, (2) the *reordering* of innate sequences, (3) the *assessment of the efficacy of behavior* on the basis of objectively verifiable information, and (4) the *internal reorganization* of stored information.

To say this in another way, primate behavior is not composed of distinct levels of cultural and instinctive mechanisms, as instinct theorists and social anthropologists both commonly believe, but of the behavioral products of a constant and functionally unitary interaction between innate and environmental information. What characterizes mammals generally and primates in particular is the capacity to forge innate behavioral components and innate releasing stimuli into novel and well integrated functional complexes, controlled by the refined sensory and motor capacities of the expanded neocortex. The distinction between learned and innate is not applicable to primates; but, as primate evolution proceeds, this distinction becomes increasingly *irrelevant* as a characterization of the *functional* organization of primate action. The thrust of primate behavioral evolution is the development of distinct modalities of action, themselves innate, that merge innate and environmental information into a common product.

If this perspective is correct, then mammals, and especially the higher primates, should exhibit distinct developmental states in which acquired and innate information can become coordinated in a common product. At least two such states are in fact known, with the possibility of further states being recognized as these phenomena are more generally studied. The two known states that appear to fit this theoretical requirement are *play* (Bekoff 1974; Bruner et al. 1976; Fedigan 1972; Freeman and Alcock 1973; Gwinner 1966; Hinton and Dunn 1967; Leresche 1976; Loizos 1967; Muller-Schwarze 1971; Owens 1975a, 1975b; Poole and Fish 1975; Savage and Malick 1977; Steiner 1969)[5] and *dreaming sleep* (Chase 1972; Evarts 1967; Flannigan 1973; Hartman 1970, 1973; Jouvet-Mounier et al. 1970; McNew et al. 1971), and both are characterized by the modification of behavior from its normal context and by flexibility in the resulting behavior. Moreover, both appear to be phylogenetically progressive states restricted to mammals and some birds and associated with immaturity. A young raven (*Corvus corax*) has been reported to throw pebbles in the air and catch them and also to play chase with a dog. In a group of handreared ravens, group specific play patterns develop, indicating a social interaction effect. The play itself consists of innate behavior patterns that occur in idiosyncratic sequences. This raven play is comparable to the play reported for mammals, in which innate behavior patterns, particularly drawn from agonistic behavior, occur in sequences that differ from normal consummatory behavior. Play fighting was found to be the most common play in mongooses, but prey-catching behaviors directed to inanimate objects also occur, as well as sexual posturing. In deer, the motor patterns of play are derived from fighting, escape, sexual, and reclining behavior, and the motor patterns of different instinctive systems can combine, indicating that the hierarchical organization of instrumental activity is missing. In rats, playful behaviors are primarily derived from agonistic behavior, and 81 percent of social play patterns are variants of adult behaviors. However, the sequential ordering of these behaviors differs from the adult context. In dogs, play bouts usually involve wrestling, jaw wrestling, and inhibited biting. In a film study of polecat play, Poole concluded that, although all the agonistic patterns occur in play, play is not agonistic behavior because actual biting is much reduced. The motor component is frequent, but the threshold of actual biting appears to be greatly elevated. In ground squirrels (*Spermophilus columbianus*), play includes approach and avoidance behaviors, nudging, rearing on the hind legs, grappling, side display, wrestling, and mounting. All of these behavior patterns also occur in adult sexual and agonistic contexts, but the transitions between one class of activity and the next may be abrupt in play. Similar phenomena have also been widely reported from primates. Rather anecdotal data also suggest that play is more elaborated in those species that are both intelligent and relatively secure, such as carnivores and primates. This lability of play is well known, and it has often been suggested, though not yet proved, that play is functionally related to the programming of novel behavioral combinations and the integration of behavior with experience (Dohlinow and Bishop 1970). However, it has not been generally realized that play shows close morphological similarities to the state of dreaming sleep, which is also characteristic of mammals and birds.

Sigmund Freud thought there were processes analogous to linguistic productivity in the ontogeny of instinctive behavior, and he had the prescience to look for instinctive productivity in dreaming sleep (1976). Since Freud's time, it has been established that dreaming in man is correlated with awakenings from the sleep state variously called dreaming, rapid eye movement, REM, paradoxical, or active sleep, and that this sleep state is physiologically distinct from slow wave sleep. The most interesting finding to emerge from the physiologist's couch was the fact that active sleep was the deepest sleep state as measured by the behavioral criteria of refractoriness to external stimuli and by the physiological necessity of passing through other sleep stages to reach it, yet it was characterized by the EEG of alert wakefulness. Electrophysiological studies of single neurons in the visual cortex also showed that these neurons could be as active in dreaming sleep as during wakefulness. It is also well known that dreaming sleep is not simply the absence of wakefulness because it is associated with the conscious recall of dreams.

With the exception of the adult egg-laying mammal *Echidna,* two distinct sleep states probably occur in all mammals, and two similar states have also been reported for birds (Chase 1972). Reptiles, however, have only an undifferentiated sleep state more similar to slow-wave, indicating that the differentiation of sleep states is characteristic of the endothermic vertebrates. Moreover, the elaboration of active sleep, measured by the relative proportions of the two sleep states during sleep periods, reaches its highest level in mammals. Active sleep is also more pronounced in the young. In humans, up to 80 percent of the sleep of neonates is active sleep, compared to about 20 percent in adults, and chimpanzees have a similar sleep cycle to man. Active sleep precedes slow-wave sleep during postnatal development in cats and rats; but guinea pigs, which are born very mature, have both kinds of sleep at birth. However, the active sleep of newborn guinea pigs is still twice adult frequencies. Related to this is the fact that active sleep is a low-priority state that exists in a buffered context. Allison and Cicchetti found that active sleep is negatively correlated with the danger of predation, with easily preyed upon species showing less REM (1976). The association of active sleep with immaturity suggests that it has some relationship to the maturation of the brain, and the verbally reported data from humans indicates that dreams have a highly charged affective tone with visual imagery predominating. The plot usually has an associational structure, which was characterized in detail by Freud himself. It is not known, of course, whether other animals share these subjective phenomena, but no major species differences in the physiology of active sleep have been reported for primates. Some behavior is characteristic of the active sleep state, such as rapid eye movements and rapid ear movements, but the spinal musculature shows a great loss of tonus, which is incompatible with behavior. These facts suggest that active sleep is a kind of wakefulness that occurs without any possibility of sensation or behavior. This characterization is not so anomalous if we compare active sleep to play.

Play shows a similar phyletic distribution to the differentiated sleep states, and it also is characteristic of young animals. Also in accord with the buffered context of both social play and active sleep is the fact that in both states the normal ordering of instinctive components is disrupted and replaced by a lability of transitions among instinctive systems. Observers of both social play behavior and dream reports have emphasized this "productive" aspect of the two states. Where play and active sleep differ, they differ in a complementary way. Play, for example, incorporates the motor patterns of emotional consummatory states, such as sex and fear, but not the species-specific signals that accompany these emotions in the nonplay context. In three hundred hours of field observation of play in rhesus monkeys, supplemented by film analysis, Symons failed to find any examples of agonistic signals (1974).[6] In fact, the occurrence of such signals usually heralds the cessation of play. In many primate species, a squeal or a vocal threat during a play bout will provoke intervention by an adult, and play is often regarded as grading into real agonism as animals get older precisely because such nonplay emotional signals start to appear. This suggests that the affect of play does not consist of the emotions whose behavior patterns occur during the play bout but rather that these behaviors are lacking in the normal affect. In dreams, in contrast, the affective monitoring of recalled events is uppermost, with events becoming distorted to accord with the prevailing emotional interpretation. But perhaps there is still more symmetry than complementarity: it is well known that penile erections are correlated with active sleep. Perhaps active sleep drops out the appetitive behavior related to the somatic musculature, and play drops out the consummatory behavior related to the visceroautonomic system.

As argued by Reynolds (1976), it is useful to view play as simulated activity; and modern technology provides many instances of simulation studies that attempt to assess the consequences of action without also engaging the life and death consequences of real action. A simulated space mission, for example, provides information that can be used to plan actual missions but does not entail the risk of lives and of many millions of dollars. Contrary to a "practice" theory, however, simulation is not simply a way of practicing what is already known but a way of integrating what is known with what it is necessary to discover. Simulation is a way of assessing the consequences of action in contexts in which mistakes are not harmful.

Transposing this argument to the behavior of animals, a simulated context could be created by inhibiting the normal outputs and inputs to neural systems. In social play, the somatic behavior patterns of instinctive systems are uncoupled from their normal consummatory acts through heightened thresholds, such as the incomplete bite of the play bite. In this way the external consequences of affective action relayed through the exteroceptive sensory systems could be assessed without direct engagement of the core brain reinforcement mechanisms that normally accompany such action. In active sleep, the somatic musculature and exteroceptors are themselves inhibited, and information from past experience can be integrated with the affective consequences of action without the motivation of overt behavior that normally accompanies such integration. The innate networks of sensory-motor connectivity, which have been revealed by electrostimulation of the core brain, can be reorganized into composite networks that integrate these innate components with environmentally acquired information about the consequences of action. Contrary to Lorenzian instincts, where the consummatory act is performed without respect to its consequences and fixed-action patterns are triggered off by innate releasers, in primates at least, these components are highly contextualized through developmental processes of selective inhibition. In chapter 3, some of the evidence for the social learning of innate responses will be reviewed. For the present, social play and active sleep can be thought of as complementary simulative states that are involved in the contextualization of affect and the selective engagement of specific innate behavioral components. The play and dream states, and the maturational factors to which they are ultimately related, make possible different adult social systems through the imposition of different inhibitory or contextualization grids onto basically similar networks of innate motor programs and affective connectivity.

To recapitulate, the structural connectivity of behavioral networks, which is genetically controlled in primitive species, is modified during genetically differentiated "liminal" states in mammals. This formulation is given credence by the fact that a number of investigations attempting to characterize mammalian behavior have arrived at a similar conclusion from very disparate starting points. Leyhausen, reflecting on his twenty thousand instances of prey catching in cats, rejected the concept of a prey-catching instinct but postulated a number of separate instinctive acts, like stalking, pouncing, and neck biting, each with their own action-specific energy. These acts, which appear in cats raised in social isolation, "are not always all united in an inflexible chain, even when the appetence for killing or eating dominates; some may be missed out or they may be mutually substituted for one another, according to circumstances" (1973a, p. 227). He concludes that "individual instincts of prey-catching have probably evolved as a result of the single links of a chain of instinctive movements originally dependent on one single source of action-specific energy becoming emancipated, growing more highly differentiated in the process, and each acquiring action-specific energy of its own. . . . Perhaps it is precisely this kind of 'breakup of instinct' where, as with the Hydra of Lerna, seven new heads (relatively independent instincts) grow when one is cut off, the most important factor in the whole of mammalian evolution" (1973a, p. 237). However, Leyhausen is careful to point out that man, who has carried this process furthest, is not therefore free to recombine these components in any possible way. Rather "the rules governing changes in the combination of units depend largely, though not exclusively, on the autonomous rhythmicity of the individual propensities and on their regulative relationships to one another, and it is *this which is the essence of the relative hierarchy of moods.*" (1973a, p. 239; italics in original.) In an analogous conclusion, the Scheibels, assessing the implications of their electrophysiological research on neurons of the reticular formation, 80 percent of which exhibited an endogenous and apparently idiosyncratic rhythmicity in their sensitivity to discrete categories of stimulation, postulated a matrix of *modular processing units*, each of which could assume pace-making control as a function of biological urgency (1967, pp. 586–587).

Leyhausen, working from observational data, and the Scheibels, considering the neural mechanisms involved in the organization of part behaviors, have both converged on the concept of innate behavioral components that function as "modules" rather than as stereotypic chains, and this perspective requires the evolution of specific mechanisms for the integration of such modules into instrumentally useful combinations. This approach is supportive of the concept of play and dreaming sleep as "modalities of recombination," which integrate environmental information with innate networks in the course of maturation. However, the complementarity of play and dreaming sleep, with the appetitive behaviors activated in the former and the consummatory behaviors in the latter, suggest a relationship to a second polarity that contrasts the internal world of homeostasis with the external world of "objective" action.

THE ELABORATION OF FUNCTION

Pribram has argued that the evolution of the primate brain can be thought of as the progressive elaboration of cognitive mechanisms to assess both *internal and external consequences of action* (in press; 1971). This functional approach crosscuts the traditional distinctions between phylogenetically older and more recent structures. Both functions are neocortically represented, both are functionally interlocked, both can call upon learned and innate behavior, and in man, at least, both are active simultaneously. These two functions can be termed here the *affective modality of action* and the *instrumental modality of action,* which regulate and evaluate internal and external consequences respectively. The instrumental and affective modalities of action have a great deal of support from neuropsychological evidence, but they cannot be equated with particular parts of the brain because they are functional characterizations. In a physiological analogy, the maintenance of constant body temperature, thermoregulation, is a functional characterization; and it involves the presence of hair on the skin, regulation of peripheral blood flow, shivering, motivation to move to another location, altered metabolic rate, perspiration, and so on. Thermoregulation is a function that incorporates diverse levels of behavior from autonomically induced movement of a single hair to a plan to insulate the house. The instrumental and affective modalities of action also embrace the entire spectrum of behavior, but like thermoregulation, some structures are more central to their respective functions than others. The affective modality is highly dependent upon the rostral parts of frontal and temporal cortex and upon the limbic structures, such as hippocampus, septum, amygdala, and hypothalamus, involved with physiological homeostasis, reinforcement, inhibition, attention, and innate behavior patterns. As Pribram has stressed, these are by no means "primitive" activities, and he has delineated three dimensions of function that characterize the phylogenetically older limbic system, basal ganglia, and core brain mechanisms: (1) a *stabile* function, which regulates physiological states of body and brain at homeostatic levels; (2) a *protocritic* function, which mediates diffusely projecting sensory systems, like pain or temperature, in which the information is evaluated in terms of urgency rather than in discrete localization in space and time; and (3) an *effective* function, which (*a*) activates the cortex in reference to significant, unexpected, or sustained stimuli, such as those functions mediated through the diffuse thalamic projection system and core brain neurochemical pathways, or which (*b*) controls the basal ganglia motor systems, with their associated innate behaviors and regulation of sensory input.

In contrast, the instrumental modality of action, which assesses the external consequences of behavior, requires the cortex surrounding the major fissures — the sensory and motor cortex and related "association" areas. These structures subserve an *epicritic* function that localizes stimulation in precise points of space-time and abstracts from experience *context-free patterns* that are not dependent upon the internal state of the animal (though presupposing wakefulness). The affective and instrumental modalities, which are functional characterizations of action, differ in some important respects from the traditional oppositions between reason and emotion. For example, a related conceptualization, Paul MacLean's, which has been widely cited in discussions of human ethology, emphasizes that the regulation of the internal milieu is a primitive function, mediated by phylogenetically primitive parts of the brain, upon which neocortical control is superimposed.[7] Pribram's conceptualization, in contrast, makes far more behavioral sense, for it postulates the involvement of phylogenetically advanced mechanisms in the assessment of both the internal and external consequences of action and requires mechanisms for the integration of information in distinct functional modalities. From this perspective, the relationship between reason and emotion, to use traditional labels, is not one of hierarchy but of *specialization by function,* and the human brain has as one of its major tasks the integration of such different kinds of information into a unified course of action. This conception of the relationship between reason and emotion as one of integration rather than of subordination cannot be overstressed, for it constitutes one of the great liberating concepts of ethological science, a concept that stands in marked opposition to the major emphasis of Western thought. The distinction between the affective and instrumental modalities of action, which are specialized for the assessment of internal and external consequences of action respectively, has great relevance to primate evolution, but as Pribram points out, the conception just related is still too simple. First, affectively significant stimuli must be transmitted and acted upon, using the external sensory and motor mechanisms of instrumental action, and, conversely, instrumental action creates internal consequences that activate affective mechanisms. Consequently, the information processing of each modality, although performing a specialized function, must nonetheless have access to the results of the other. For this reason, a behavioral theory of primates that studies each system as if it were functionally autonomous is not as likely to produce as interesting conclusions as will a theory that stresses *both* functionally specialized behavioral systems *and* the interaction of functionally specialized systems. Affective behavior (discussed in chapter 3), for example, does not exist as a specific level, but rather the specialized sensory and motor components of these mechanisms are *interlocked* with the sensory and motor mechanisms of external assessment mediated by instrumentality. Higher primates, in other words, through the differentiation of the liminal states of play and dreaming sleep, have evolved a capacity for *composite networks* that interrelate innate perceptual and motor mechanisms specialized for affectivity with perceptual and motor mechanisms specialized for instrumentality. Second, a concept of affectivity as "internal assessment" is too restrictive, for it is quite evident that affective systems are also *specialized for the media-tion of social relationships*. Again, in contradistinction to the weight of the Western intellectual tradition, affect is not to be thought of as a flux of private and idiosyncratic states but as *the phylogenetically specialized channel by which internal states are made public*. Viewed in this way, it is quite clear why affective communication in man does not lapse into disuse, as the Ladder of Reason would predict it would, and it argues strongly against the view that such communication is anthropologically uninteresting simply because related forms occur in animals. Quite the contrary, it suggests that affective communication should differentiate with the internal states them-selves. Moreover, an interlocking of affective display and conceptually en-coded information, suggested by the example of *parry,* discussed above, would lead one to expect conceptual modulation of the resulting display.

The theoretical constructs of affective and instrumental modalities, conceived of as specialized systems that interact, provide a potentially useful model for primate behavioral evolution. Contrary to nineteenth century evolutionism, with successive emergence of levels, an ethological perspec-tive argues for a *progressive development* of *the constituent mechanisms involved in affective-instrumental integration.* In chapter 3, for example, the evidence for the progressive elaboration of the ability for self-recognition in primates is summarized, and the present approach would predict systematic differences in the presentation of self in the social interactions of different species of primates. The affective system of apes would have access to a self-concept that included the ability to recognize oneself in a mirror, whereas the mon-key's affective system would not. Human affectivity, on the basis of further changes in instrumental action, discussed in chapter 4, would presumably be able to employ a self-concept that included the hypothetical images of oneself that are held by others. Monkeys, apes, and humans would there-fore be expected to differ on their relative abilities in the presentation of self, and differences in the social integration processes would follow from that. Yet one would need only to postulate differential conceptual processing of the body image in primates and not to propose any major transformation of affective dynamics per se.

Grooming behavior, dealt with in chapter 5, provides a second illus-tration. On the basis of comparative data derived from studies of object using, the case is made that there is a phylogenetic progression in construc-tional abilities and instrumental action in the primate lineage leading to man. Because constructional action approximates a progressive evolution-ary series, activities that employ motor programs for dextrous manipula-tion ought to change their behavioral form as human evolution proceeds. The behavior should change from dextrous manipulation of objects to the creation of enduring and novel object constructions. Nonetheless, these evolutionary elaborations retain the motivational connections of the more primitive forms and continue to operate in accordance with the same affective dynamics, as suggested by the similarity between grooming behavior in monkeys and object exchange in man.

To summarize this argument, the integration of instrumentality and affectivity shows a great deal of phylogenetic continuity in primates, as do the constituent affective components. However, the comparative evidence also supports progressive changes in conceptualization, instrumental skills, and the volitional control of behavior in the course of human evolution. These changes give rise to *qualitatively different systems of social coordination* that nonetheless continue to be implemented and assessed by those affective mechanisms specialized for the public display of internal state and for the motivation of socially directed behavior. In other words, primate institutions—the product of coordinated social action—are both behaviorally discontinuous and psychologically comparable.

The products of social interaction differ in man, but this change in the power of institutions is not paralleled by comparable discontinuities at the psychobiological level. Also, as discussed in chapter 6, human language, although a phylogenetically new form of social coordination species-specific for man, can nonetheless be explained as the further elaboration of affective-instrumental integration, involving the conjunction of innately controlled vocal effectors and conceptually encoded information. In chapter 7, a summary of some of the behavioral differences between man and other primates is provided, and the suggestion is made that many of the most divergent aspects of the human biogram—the use of internal imagery to control constructional activity and the power of words to motivate human action—are part of an adaptation based on functional hallucination, itself created by minor genetic changes.

NOTES TO CHAPTER 2

1. For the use of classical ethological concepts in the explanation of behavior, see Tinbergen (1950, 1951); Lorenz (1957, 1973); Ewer (1968); Darwin (1872, 1873); and E. Hess (1962, 1973). The critique of classical ethology can be found in Hinde (1970, pp. 407, 418–419); Rowell (1961); Andrew (1956a, 1956b); Leyhausen (1973).

2. See also Huxley (1966); Morris (1956, 1957, 1966); Tembrock (1968); Daanje (1951); Blest (1961); and Andrew (1956–1957) for innate ritualization.

3. Also see chap. 3 of this book.

4. Also see references in chaps. 1, 6.

5. Also see chap. 3 of this book and the papers by Lawick-Goodall, Sade, Bertrand, and Loizos in Bruner et al. (1976). For a review of the postnatal changes in the brain, see Trevarthen (1973).

6. See also Bertrand in Bruner et al. (1976).

7. See n. 55, chap. 1.

REFERENCES

AKERT, K. 1961. Diencephalon. In *Electrical stimulation of the brain,* ed. D. E. Sheer, pp. 288–310. Austin: U. of Texas Press.

ALLISON, TRUETT, and CICCHETTI, DOMENIC V. 1976. Sleep in mammals: ecological and constitutional correlates. *Science* 194:732–734.

ANDREW, R. J. 1956a. Some remarks on behaviour in conflict situations, with special reference to *Emberiza* spp. *Brit. J. Anim. Behav.* 4:41–45.

1956b. Normal and irrelevant toilet behavior in *Emberiza* spp. *Brit. J. Anim. Behav.* 4:85–91.

1956–1957. Intention movements of flight in certain passerines and their uses in systematics. *Behaviour* 10:179–204.

1972. The information potentially available in mammal displays. In *Nonverbal communication,* ed. R. A. Hinde, pp. 179–206. Cambridge: Cambridge U. Press.

BARD, PHILIP, and MACHT, MARTIN B. 1958. The behaviour of chronically decerebrate cats. *CIBA Foundation Symposium on the Neurological Basis of Behaviour,* pp. 55–75. London: Churchill.

BARD, PHILIP, and MOUNTCASTLE, VERNON B. 1948. Some forebrain mechanisms involved in the expression of rage with special reference to angry behavior. *Research Publications of the Assoc. in Nervous and Mental Disease* 27:362–404.

BARLOW, GEORGE W. 1968. Ethological units of behavior. In *The central nervous system and fish behavior,* ed. David Ingle, pp. 217–232. Chicago and London: U. of Chicago Press.

BEKOFF, MARC 1974. Social Play in Canids. *American Zoologist* 41:323–340.

BLEST, A. D. 1961. The concept of ritualization. In *Current problems in animal behaviour,* ed. W. H. Thorpe and O. L. Zangwill, pp. 102–124. Cambridge: Cambridge U. Press.

BRUNER, J. S.; JOLLY, A.; and SYLVA, K., eds. 1976. *Play: its role in development and evolution.* Harmondsworth, Eng.: Penguin Books.

CHASE, M. H., ed. 1972. *The sleeping brain. Perspectives in brain science 1.* Los Angeles: Brain Information Service/Brain Research Institute.

DAANJE, A. 1951. On locomotory movements of birds and the intention movements derived from them. *Behaviour* 3:48–98.

DARWIN, CHARLES   1872. *The expression of the emotions of man and animals.* London: Murray.

  1873. Origin of certain instincts. *Nature. A weekly illustrated journal of science* 7:417– 418. Reprinted in *The collected papers of Charles Darwin,* vol. 2, ed. Paul H. Barrett, pp. 172– 176. Chicago: U. of Chicago Press.

DOLHINOW, P. J., and BISHOP, N.   1970. The development of motor skills and social relationships among primates through play. *Minnesota Symposium on Child Psychology 6.* Minneapolis: U. of Minnesota Press.

EIBL-EIBESFELDT, IRENAUS   1975. *Ethology: the biology of behavior.* 2nd ed. New York: Holt, Rinehart and Winston.

EVARTS, EDWARD V.   1967. Unit activity in sleep and wakefulness. In *The neurosciences: a study program,* ed. G. C. Gardner, T. Melnechuk, and F. O. Schmitt, pp. 545– 556. New York: Rockefeller U. Press.

EWER, R. F.   1968. *The ethology of mammals.* New York: Plenum.

FEDIGAN, LINDA   1972. Social and solitary play in a colony of vervet monkeys (*Cercopithecus aethiops*). *Primates* 13:347– 364.

FLANIGAN, W. F. Jr.   1973. Sleep and wakefulness in Iguanid lizards, *Ctenosaura pectinata* and *Iguana iguana. Brain, Behavior, and Evolution* 8:401– 436.

FREEMAN, H. E. and ALCOCK, J.   1973. Play behaviour of a mixed group of juvenile gorillas and orang-utans *Gorilla g. gorilla* and *Pongo p. pygmaeus. International Zool. Ybk. 13,* pp. 189– 194.

FREUD, SIGMUND   1976. *The interpretation of dreams.* Trans. James Strachey. Hammondsworth, Eng.: Penguin Books. 1st German ed., 1900.

GWINNER, EBERHARD   1966. Uber einige Bewungsspiele des Kolkraben (*Corvus corax* L.). *Z. Tierpsychol.* 23:28–36.

HARTMANN, ERNEST L.   1973. *The function of sleep.* New Haven: Yale U. Press.

HARTMANN, ERNEST, ed.   1970. *Sleep and dreaming.* Boston: Little, Brown.

HESS, E. H.   1962. Ethology: an approach toward the complete analysis of behavior. In *New directions in psychology,* ed. Brown, Hess, and Mandler, pp. 157– 226. New York: Holt, Rinehart and Winston.

  1973. *Imprinting: early experience and the developmental psychobiology of attachment.* New York: Van Nostrand.

HESS, W. R.   1957. *The functional organization of the diencephalon.* New York: Grune and Stratton.

HINDE, R. A.   1953. Appetitive behaviour, consummatory act, and the hierarchical organisation of behaviour— with special reference to the great tit *(Parus major). Behaviour* 5:189– 224.

  1956. Ethological models and the concept of drive. *Brit. J. Philosophy of Science* 6:321– 331.

  1959. Unitary drives. *Animal Behaviour* 7:130– 141.

  1970. *Animal Behaviour: a synthesis of ethology and comparative psychology.* New York: McGraw-Hill.

HINTON, H. E., and DUNN, A. M. S. 1967. *Mongooses: their natural history and behaviour.* Berkeley and Los Angeles: U. of California Press.

HUXLEY, JULIAN 1923. Courtship activities in the red-throated diver *(Colymbus stellatus Pontopp);* together with discussion of the evolution of courtship in birds. *Linnean Soc. Proc.* 53:253–292.

1966. Introduction, a discussion of ritualization of behaviour in animals and man. *J. Royal Soc.* 251, no. 772:249–271.

JOUVET-MOUNIER, DANIELE; ASTIC, L.; and LACOTE, D. 1970. Ontogenesis of the states of sleep in rat, cat, and guinea pig during the first postnatal month. *Developmental Psychobiology* 2:216–239.

KENNEDY, J. S. 1954. Is modern ethology objective? *Brit. J. Anim. Behav.* 2:12–19.

KORNHUBER, H. H. 1974. Cerebral cortex, cerebellum and basal ganglia: an introduction to their motor functions. In *The neurosciences: third study program,* ed. F. O. Schmitt and F. G. Worden, pp. 267–280. Cambridge: MIT Press.

LEHRMAN, DANIEL S. 1953. A critique of Konrad Lorenz's theory of instinctive behavior. *Quarterly Review of Biology* 28:337–363.

LERESCHE, LINDA A. 1976. Dyadic play in hamadryas baboons. *Behaviour* 57:190–205.

LEYHAUSEN, PAUL 1973a. On the function of the relative hierarchy of moods (as exemplified by the phylogenetic and ontogenetic development of prey-catching in carnivores). In *Motivation of human and animal behavior: an ethological view,* ed. K. Lorenz and Paul Leyhausen, pp. 144–247. New York: Van Nostrand Reinhold.

1973b. Theoretical considerations in criticism of the concept of the "displacement movement." In *Motivation of human and animal behavior: an ethological view,* ed. Konrad Lorenz and Paul Leyhausen, pp. 59–69. New York: Van Nostrand Reinhold. German ed., 1952.

LOIZOS, C. 1967. Play behavior in higher primates: a review. In *Primate ethology,* ed. D. Morris, pp. 176–218. Chicago: U. of Chicago Press.

LORENZ, KONRAD 1957. The nature of instinct. Trans. Clair H. Schiller. In *Instinct: the development of a modern concept,* ed. C. H. Schiller, pp. 129–175. London: Methuen.

1958. The evolution of behavior. *Scientific American,* December.

1971. Comparative studies of the motor patterns of Anatinae. Trans. R. Martin. In *Studies of human and animal behaviour,* vol. 2, pp. 14–113. London: Methuen. 1st German ed., 1941.

1973. The comparative study of behavior. In *Motivation of human and animal behavior: an ethological view,* ed. Konrad Lorenz and Paul Leyhausen, pp. 1–31. New York: Van Nostrand. German ed., 1939.

McDOUGALL, WILLIAM 1908. *An introduction to social psychology.* London: Methuen.

McKINNEY, F. 1965. The comfort movements of Anatidae. *Behaviour* 25:120–220.

McNEW, J. J.; HOWE, R. C.; and ADEY, W. R. 1971. The sleep cycle and subcortical-cortical EEG relations in unrestrained chimpanzee. *Electroencephalography and Clinical Neurophysiology* 30:489–503.

MEARS, CLARA E. 1978. Play and the development of cosmic confidence. *Developmental Psych.* 14:371–378.

MORRIS, DESMOND 1956. The feather postures of birds and the problem of the origin of social signals. *Behaviour* 9:75–113.

1957. "Typical intensity" and its relation to the problem of ritualization. *Behaviour* 11:1–12.

1966. Abnormal rituals in stress situations: the rigidification of behaviour. *J. Royal Soc.* 251, no. 772: 327–330.

MULLER, J. 1843. *Elements of physiology.* Trans. William Baly. Arranged from the 2nd London ed. by John Bell. Philadelphia: Lea and Blanchard.

MULLER-SCHWARZE, DIETLAND 1971. Ludic behavior in young mammals. In *Brain Development and behavior,* ed. M. B. Sterman, D. J. McGinty, and A. M. Adinolfi, pp. 229–249. New York: Academic Press.

OWENS, N. W. A. 1975a. Social play behaviour in free-living baboons, *Papio anubis. Animal Behaviour* 23:387–408.

1975b. A comparison of aggressive play and aggression in free-living baboons, *Papio anubis. Animal Behaviour* 23:757–765.

POOLE, T. B. and FISH, J. 1975. An investigation of playful behaviour in *Rattus norvegicus* and *Mus musculus* (Mammalia), *J. Zoology* (London) 175:61–71.

PRIBRAM, KARL H. Forthcoming. Emotions. In *Handbook of Clinical Neuropsychology,* ed. S. B. Filskov and T. J. Boll.

REYNOLDS, PETER C. 1976. Play, language and human evolution. In *Play: its role in development and evolution,* ed. J. S. Bruner, A. Jolly, and K. Sylva, pp. 621–635. Harmondsworth, Eng.: Penguin Books.

REYNOLDS, VERNON 1976. *The biology of human action.* Oxford: Oxford U. Press.

ROWELL, C. H. F. 1961. Displacement grooming in the chaffinch. *Animal Behaviour* 9:38–63.

SAHLINS, MARSHALL 1976. *The use and abuse of biology: the anthropological critique of sociobiology.* Ann Arbor, Mich.: U. of Michigan Press.

SAVAGE, E. S., and MALICK, C. 1977. Play and socio-sexual behavior in a captive chimpanzee *(Pan troglodytes)* group. *Behaviour* 60:169–194.

SCHEIBEL, M. E., and SCHEIBEL, A. B. 1967. Anatomical basis of attention mechanisms. In *The neurosciences: a study program,* ed. G. C. Quarton,

T. Melnechuk and F. O. Schmitt, pp. 577–602. New York: Rockefeller U. Press.

SCHLEIDT, WOLFGANG M. 1974. How "fixed" is the fixed action pattern? Z. Tierpsychol. 36:184–211.

SEVENSTER, P. 1961. A causal analysis of a displacement activity (fanning in Gasterosteus aculeatus L.). Behaviour, supplement no. 9.

STEINER, A. L. 1969. Play activity of Columbian ground squirrels. Zeitschrift fur Tierpsychol. 28:247–261.

SYMONS, DONALD 1974. Aggressive play and communication in rhesus monkeys (Macaca mulatta). American Zoologist 14, no. 1: 317–322.

TEMBROCK, GÜNTER 1968. Land mammals. In Animal communication: methods of study and results of research, ed. T. A. Sebeok, pp. 338–404. Bloomington: Indiana U. Press

THORPE, W. H. 1966. Ritualization in ontogeny: I. Animal play. In Ritualization of behaviour in animals and man, ed. J. Huxley, Proc. Royal Soc., London 251 B, pp. 311–319.

TINBERGEN, N. 1950. The hierarchical organization of the nervous mechanisms underlying instinctive behaviour. In Physiological mechanisms in animal behaviour. Symposia of the Society for Experimental Biology no. 4. New York: Academic Press.

1951. The study of instinct. Oxford: Oxford U. Press.

1952. "Derived activities"; their causation, biological significance, origin and emancipation during evolution. Quarterly Rev. Biol. 27:1–32.

1959. Comparative study of the behaviour of gulls (Laridae): a progress report. Behaviour 15:1–70.

TREVARTHEN, COLWYN B. 1973. Behavioral embryology. In Handbook of perception. III. Biology and perceptual systems, ed. E. C. Carterette and M. P. Friedman, pp. 89–117. New York and London: Academic Press.

VON HOLST, ERICH, and VON SAINT PAUL, URSULA 1963. On the functional organization of drives. Animal Behaviour 11:1–20.

WICKLER, WOLFGANG 1967. Socio-sexual signals and their intra-specific imitation among primates. Primate ethology, ed. D. Morris, pp. 69–147.

WILSON, E. O. 1975. Sociobiology: the new synthesis. Cambridge, Ma.: Harvard U. Press.

3

ON THE END OF THE WAR OF
EACH AGAINST ALL

ABOUT FACE

The evolutionary continuity of the affective modality is manifested primarily by the face, and it was Darwin who began the comparative study of facial expression (Petrinovich 1973, pp. 229–230; Darwin 1965). Moreover, he introduced all the methods still in use today. He examined facial expression in children, in the insane, and in the blind. He had a man's face photographed and submitted the photos to independent judges, and he corresponded with colonialists on the facial expressions of non-Europeans. Although Darwin had met savages (as they were then known) on the voyage of the *Beagle,* an adequate crosscultural study of facial expression could not be done until the techniques of photography had substantially improved. Although the Russian primatologist L. Kohts published a series of photographs on facial expressions of the chimpanzee in 1935, there was little interest in primate expression during the interval between Darwin's book in 1872 and the upsurge in behavioral primatology in the 1960s (Chevalier-Skolnikoff 1973, p. 18; Yerkes and Yerkes 1970, chap. 22; Finch 1942; Hebb 1945, 1961). However, psychologists had presented Koht's photographs of chimpanzee facial expressions to panels of judges, and the results showed that American college students interpreted chimpanzee expressions in the same way as did the chimps' Russian primatologist-caretaker, with one exception (Foley 1935). The students misinterpreted the fear grimace or silent bared teeth display of chimpanzees as a smile, indicative of pleasure. Chimpanzees, in fact, have been induced to "smile" for the movies by the application of mildly aversive stimuli.

The judgment method was also applied crossculturally in a number of studies by different researchers, sometimes with the intent of demonstrating the arbitrariness of facial expression (Ekman 1973, chap. 4; Ekman 1977; Izard 1971). Although the methods varied, the results of these studies were consistent with one another. Humans in different countries judge drawings and photos of facial expressions made elsewhere in a manner that exhibits *both* cultural variation and panhuman agreement. Ekman and his colleagues, like the European ethologists we will discuss, distinguish between displays of a unitary emotion, characterized by a particular muscular configuration, and *blends* that display two or more emotions simultaneously.

Many psychologists, but by no means all, have argued for a small set of distinct emotions. Tomkins, basing his study on facial expression, postulated eight basic types (1962). Ekman and Friesen chose photographs that exhibited six of the eight primary emotions of Tomkins—happiness, sadness, anger, fear, surprise, and disgust—and used them as test stimuli in a crosscultural study (1971). Given lexical equivalents in their own language for these primary affects, with the addition of a seventh word, *contempt,* subjects in Japan, the United States, and three South American countries matched the pictures with the terms in the same way as the experimenters. Agreement was not perfect, but it was statistically significant, with the best agreement on happiness. In another study by Izard, the eight primary affects were used: interest-excitement, enjoyment-joy, surprise-startle, distress-anguish, disgust-contempt, anger-rage, shame-humiliation, and fear-terror. Although the terminology differs, Izard's categories, which are also based on Tomkins's classification of the primary affects, are basically similar to Ekman's list, with the addition of shame-humiliation and interest-excitement. Izard's study, however, showed less crosscultural agreement than that of Ekman and Friesen. The same uniformity was found among Europeans and Americans, but the Japanese subjects and also African students tested in Paris with French words both showed statistically significant disagreements from the experimenter's classification. In the Japanese case, Ekman later retested his own subjects with Izard's photographs but with his own set of Japanese emotion terms and obtained higher correlations. However, as Ray Birdwhistell soon pointed out, the question was not whether peoples with a common cultural experience, like Americans, Europeans, and modern Japanese, had similar facial expressions but whether this uniformity extended to different cultural traditions.

To answer this question, Ekman and Friesen undertook several studies of the Fore people in New Guinea (West Irian). The direct application of the method that worked so well on the literate subjects gave poor agreement among the Fore. However, good agreement was found by developing alternate testing procedures. In one, the subjects were told in Fore that they were to choose one of three pictures that typified a particular emotion. Thus, the Fore subject was told "she is angry and is about to fight," and was given a choice of three photographs, all used in the earlier studies of literate cultures, and one of which was an example of anger as previously judged by the literate subjects. Statistically significant agreement was found with the literate classifications on this task, except in the discrimination of fear from surprise. Ekman and Friesen also asked Fore subjects to pose various emotions. These expressions were videotaped and the tapes shown to a panel of American college students who were asked to match the expressions to the primary emotion words in English. Again, good agreement was obtained, except for surprise and fear, which the college students had difficulty in discriminating among the posed Fore expressions. Ekman has argued that much of the crosscultural variability in emotional expression is due to cultural *display rules* that enhance or inhibit particular expressions in particular contexts. In accordance with Tomkins's theory, these display rules would also be affectively mediated, with shame, for example, inhibiting crying in American adult males.

Ekman is unusual in having systematically studied both the perceptual and motor aspects of facial expression crossculturally, but some additional information on the innateness of human expressions is provided by children born both blind and deaf. These children do in fact produce a basic expressive repertory, as photographically documented by Eibl-Eibesfeldt (1973). However, blind children, since they would not be able to correct their expressions visually, would also be expected to deviate from the cultural norms in the modulation and contextualization of their innate motor acts. Anthropology, which might be expected to shed light on such interactions, is not likely to contribute much to our understanding if it persists with the assertion made by Edmund Leach, without benefit of a single study, that gestures that have "no ethological counterpart" form "the greater part of human non-verbal communication as viewed in a cultural context" (1972, p. 316). It is doubtful that this generalization would even be true for filmed social interaction at High Table, King's College, Cambridge; but even it if were, there would still be no justification for denigrating the importance of numerically less frequent ethological counterparts without systematic study of their actual effects.

Weston LaBarre, in noting that anthropologists are "wary of those who speak of an 'instinctive' gesture on the part of a human being" (1947, p. 49), failed to distinguish between the motor aspect of the gesture and the interpretation of it by others and between those gestures most closely related to affective states and those that are highly conventional. Since there is no reason to suppose that human beings could not simultaneously possess innate gestures, conventional gestures, and innate motor acts that encode conventional meanings, all coexisting in the communicative repertory, these distinctions cannot be ignored. Nonetheless, the data presented by LaBarre are in some respects more congenial to the hypothesis of crosscultural similarity than to his own. He tells us, for example, that the Copper

Eskimo welcome strangers with a buffet on the head or shoulders with a fist; Northwest Amazonians slap one another on the back; Polynesian males greet with embracing and back rubbing; Spanish-American males with stereotyped embracing and back patting; Torres Straits islanders with mutual hand touching and palm scratching; the Ainu of Yezo with hand grasping, ear grasping, and stroking of the face and shoulders; Kayan males of Borneo by grasping the forearm or embracing; Kurd males by mutual hand clasping and hand kissing; and the Andamanese sit one on the lap of the other and embrace. However, the natives of Matavi are truly exceptional: after prolonged absences, they thrust shark's teeth into their temples and induce bleeding. In all the other cases cited by LaBarre, the greeting involves contact behavior directed to either the back or to the hands. Chimpanzees, one of the few primate species to have a social organization in which members of the group are often away, also have greeting behavior. Like the human cases, chimpanzee greeting has a large tactile component (Van Lawick-Goodall 1968, pp. 364–368; Eibl-Eibesfeldt 1971). Van Lawick-Goodall records bowing, bobbing, crouching, touching, kissing, embracing, grooming, presenting, mounting, touching the genitals, and sometimes hand holding in the greeting behavior of these animals. As Paul Ekman has pointed out (1973, pp. 181–183), LaBarre's primary concern is not with the broad category of human gestures at all but with one of its aspects; namely those stylized representations of feelings called *emblems,* which are in fact culturally variable. The possibility of innate expression is not precluded by the presence of such signals, and the evidence indicates that humans typically possess both.

The motor components of facial display in man are clearly innate. There is not only the evidence from children born blind and deaf, but the smiles of identical twin infants are more concordant than those of fraternal twins (Freedman 1965). Also, there is the extensive homology between display and affective context among different species of primates. Neural pathology shows that these displays are represented in both voluntary and involuntary form in adult humans. For example, Monrad-Krohn has published photographs of a patient with a bullet wound in the left parietal-temporal-frontal region of the brain (Monrad-Krohn 1924; Myers 1969). The patient exhibited partial paralysis (paresis) on the right side of his face during voluntary facial movements, but his involuntary smiles were symmetrical. In some other cases, the involuntary facial expression is more pronounced on the partially paralyzed side, suggesting a disinhibition type of phenomenon. Conversely, in patients suffering from Parkinson's disease, there is a loss or asymmetry of involuntary facial expression, but willed movements of the affected parts of the face are normal (Poeck 1969). Also, in pseudobulbar palsy, there is complete paralysis of the voluntary movements of swallowing, articulation, and chewing, but the displays of laughing and crying are normal.

The presence of innate fixed-action patterns in primate displays and the association of particular patterns with particular affective states allows ethologists to trace the homologies among facial expressions. The technique is to compare facial gestures that are morphologically similar in different primate species and to interrelate these gestures to other kinds of associated behaviors. An affective-state approach would look for consistent interrelationships among the various behavior patterns and gestures of a single individual. For example, if a frown is expressive of anger, it ought to be associated in time with other behaviors indicative of anger, such as biting and hitting. A communicative approach to facial expression, on the other hand, would look for temporal relationships between the expression of one individual and the behavior of others. The social transitions may be quite different from the intraindividual transitions, as there are many possible responses to an expression by someone else. However, an affective-state theory would predict that these responses would be affectively consistent within each individual responder: that an affect, once aroused, would constrain subsequent behavioral transitions within that same individual. The arguments about the primacy of affective-state models over communication models of facial display, or vice versa, have impaired the quality of descriptive research. There are no studies that have combined both intraindividual and interindividual behavioral transitions, although an integrated approach to facial expression would argue that facial display is both the product of affect *and* a social signal interpreted by others. Moreover, neither framework in isolation can explain the behavior: facial gestures are communicative precisely because they are reliably related to what the sender is likely to do next and they are particularly salient because they activate the affective systems of the receivers. Both of these connections have innate components. Yet if there were no receivers, there would be no expression of the emotions at all; and if the signals worked simply as innate releasing mechanisms, then primate behavior would be much more predictable than in fact it is. It is this juxtaposition that a human ethology must reconcile, and the concept of central affective states is of critical importance in such a theory.

Ethologists have also used the concept of central emotional states to bring order to the communicative repertories of different species of primates. Van Hooff, for example, has systematically compared the facial displays of different primate species to establish homologies within the order (Van Hooff 1967, 1972, 1973; Redican 1975, pp. 128–137). As Darwin maintained, comparison of man with the other primates has demonstrated numerous homologies, but some difficulties have been found in tracing the morphology of the human smile and the human laugh. The smile, with retraction of the lips at the corners and variable exposure of the teeth, is most similar in morphology to the *bared-teeth face* of nonhuman primates. This display in turn has two forms. The first is a *vocal* version, found in all primates and most mammals, in which the expression is accompanied by a high-pitched vocalization, such as a scream. The *silent* version is less widespread, but it occurs in the Old World monkey genera *Macaca, Theropithecus, Mandrillus,* and *Papio* and in the chimpanzee, *Pan.* Both the silent and vocal versions are characterized by full retraction of the mouth corners and lips, exposing the teeth and gums, with the mouth itself only slightly open and the eyes open either widely or normally. The bared-teeth display is widely known among primatologists as the fear grimace, and, as the name suggests, it is normally regarded as associated with the affective state of fear. However, as Darwin (1965, p. 133) observed in the Celebes monkey *Cynopithecus* (now called *Macaca nigra*) (Van Hooff 1972, pp. 215–216), the silent bared-teeth display is sometimes given in affiliative situations by some nonhuman species, making it more comparable to the human smile. In the gelada monkey, *Theropithecus,* it appears to have evolved into a "lip-flip" display, in which the upper lip is retracted to reveal the pink mucosa of the underside of the lip. The association of the vocal bared-teeth display with fear, even terror, and the widespread acceptance of the silent display as indicative of fear in many species, as well as apparent convergence in *Macaca nigra* to an affiliative gesture, raises some interesting problems for the phylogeny of the human smile. Moreover, there is another problem with the smile, which becomes apparent when laughter is considered, as laughter and smiling can overlap.

The bared-teeth displays coexist in primate species with another display that is similar in appearance, the play-face or *relaxed open-mouth display* (Van Hooff 1972, p. 217; Redican 1975, pp. 153–156; Chevalier-Skolnikoff 1974). In this expression, the mouth is open wide but the teeth remain largely covered by the lips. The corners of the mouth distinguish it clearly from the bared-teeth displays: in the latter, the corners are pulled back and out, giving the face a more smile-like configuration. The relaxed open-mouth display is characteristic of social play, and it occurs in all apes and probably in all cercopithecine monkeys. Since mock biting is the normal activity in the rough-and-tumble play of mammals, Van Hooff has interpreted the play face as a ritualized intention movement of biting. A similar expression, with open mouth and completely covered teeth, has also been described and photographed for children. In apes, the play-face can be elicited by tickling, and human children usually produce a *wide-mouth laugh* in this context that is morphologically similar to both the play-face and the fear-grin. In chimpanzees the play-face is also accompanied by a vocalization described as staccato rhythmic breathing and rhythmic low-pitched grunting, not unlike laughing. Tomkins does not recognize play as a primary affect, and his description of anger-rage hardly conforms to the play situation: "Frown, clenched jaw, red face" (1962, vol. 1, p. 337). Rather, the facial expressions of play are closer to his affect system of enjoyment-joy: "Smile, lips widened up and out." Since Bateson's observations (1955), it has generally been appreciated that mammalian play is paradoxical, because it contains the behavioral components of aggressive behavior such as biting, hitting, scratching, butting, grappling, and chasing, with an affective component that is manifestly joyful rather than agonistic. Van Hooff's investigations into the motivational correlates of chimpanzee display confirm the association of the play-face with an enjoyable state, in spite of the agonistic form of the play behavior itself.

Using a repertory of behavior patterns, Van Hooff recorded the sequence of behaviors given by *single* chimpanzees within a social interaction (1973), where an "interaction" was defined as any period of social activity delimited by either periods of nonactivity or solitary activity of ten to twenty seconds duration or by a change in social partner. By arranging all the behavior patterns in a matrix, it is possible to compute the transitional probability of one behavior pattern being preceded or succeeded by another in the *same* animal. Although Dawkins (1976) has argued that cluster analysis should be based ultimately on the "substitutability" of behavior patterns, which classes together those that can occur in the same temporal slot, rather than just those in temporal proximity, the latter method is nonetheless revealing. The actual number of transitions from one behavior pattern to another is compared to the number that would be expected if the transitions were random. These data reveal that certain behavior patterns cluster together within the action of individuals. The relaxed open mouth display is central to a behavioral complex that includes hand-wrestle, gnaw-wrestle (play bite with grappling), gnaw (play bite), pull on another's limb, grasp, poke, gymnastics, hit, and gallop. The element *hit* connects play behavior to actual attack behavior, including trample, tug, charging, and bite. *Hit* also connects to a threat complex centered around the *arm-sway* display, comprising the bipedal swagger and bipedal arm waving described by Van Lawick-Goodall (1968, pp. 318–319). In contrast, the vocal bared-teeth display forms part of a fear complex that includes crouching, flight, avoidance, and hesitant approach. It is not accessible directly from the play-face complex except through the behavior pattern *parry,* which is raising an arm in front of the head to shield it from a blow. These cluster analysis data therefore provide evidence for a play complex that includes the relaxed open-mouth face and a fear complex that includes the vocal bared-teeth face.

Further analysis of these data, using factor analysis, distinguished seven principle factors. The factors distinguish behavioral groupings which Van Hooff termed affinitive (affiliative), play, aggression (attack and threat), submission (fear), excitement, comfort (grooming), and "show," which includes vertical head nodding, copulation, and elements of aggressive display. In this analysis, the two bared-teeth displays can be clearly distinguished. The *silent* bared-teeth display is part of the affinitive group, with the *vocal* bared-teeth display part of submission. Additional component analysis designed to further differentiate the affinitive group revealed the latter to be composed of a grooming component, a juvenile behavior component, a male sexual component, a presenting component, and a silent bared-teeth display component. The last included to a lesser extent the bared-teeth yelp vocalization, crouch-presenting, and embrace—all of which are also stronger members of other components. These analyses indicate a functional difference between the silent and vocal forms of the bared-teeth display, with the vocal version strongly related to fear, and the silent display associated with diverse affiliative behaviors. The silent bared-teeth display connects to the juvenile behavior component through the bared-teeth yelp, to presenting through the crouch-present, and to grooming through to embrace.

Van Hooff's analysis provides good support for the concept of primate display as systematically related to a small repertory of emotions, and it is possible to make a partial match between the primary affects used in the crosscultural investigation of facial expression and the components that have been delineated in the behavior of chimpanzees. Tomkins's list of eight primary affects in man—interest-excitement, enjoyment-joy, surprise-startle, distress-anguish, fear-terror, shame-humiliation, contempt-disgust, and anger-rage—has some direct parallels in chimpanzees. The aggression and fear components have a long history in ethology and clearly match anger-rage and fear-terror. Tomkins's interest-excitement primary affect is characterized as "eyebrows down, track, look, listen." Van Hooff's usage of this term refers primarily to the rising-hoot display, which is one of the most dramatic to be found in primates. Van Lawick-Goodall includes it among displays "not obviously aggressive" and lists intergroup contact situations, branch waving, and meat eating among the contexts in which it occurs. All of these situations are interesting and exciting to the participants and observers. The close affinity of locomotor and aggressive behaviors to the relaxed open-mouth face suggests a distinct play system, as most discussions of play have suggested, although Tomkins's closest analog would be enjoyment-joy. Tomkins's shame-humiliation primary affect, characterized by "eyes down, head down" perhaps has an analog in those chimpanzee behaviors that are both affinitive and submissive, such as the silent pout, pout moan, stretch pout whimper, and the silent bared-teeth display. Van Lawick-Goodall also reports whimpering and crying by chimpanzee children that are lost or are having tantrums, and these displays probably correspond to what Tomkins has called distress-anguish: "Cry, arched eyebrows, mouth down, tears, rhythmic sobbing." Tomkins's contempt-disgust affect does not match a chimpanzee display, but it bears an obvious similarity to protective withdrawal responses evoked by sudden contrastive stimuli, which are discussed by Andrew (1964, pp. 280–283). Tomkins's surprise-startle is characterized by "eyebrows up, eye blink," and Eibl-Eibesfeldt has filmed a sudden eyebrow raise as part of human greeting in diverse cultures (1972, p. 19). However, there are no comparable data on chimpanzees. Nor are there parallels in Tomkins's primary affect list to sexuality, juvenile behavior, or groom and contact behavior, even though all of these are generally recognized as useful categories by primatologists and all have homologous forms in man. Sex and grooming crosscut the affective systems as Tomkins defines them because they are affective but lack the distinctive facial component. However, they have distinctive somatic components, suggesting that Tomkins's list is in fact too narrow. Play, as has already been seen, is likely to be a distinct state of neurochemical modulation with maturational changes, so motivationally it also appears equivalent to affects. Also, it does have distinct facial expressions. However, as Van Hooff has pointed out, there is still a major difference between the emotional displays of human beings and chimpanzees: both species have expressions which are morphologically similar to the play open-mouth display and the silent bared-teeth display, but only humans smile and laugh. There appears to be a functional change in these two display patterns in man when compared to other primates that has raised great difficulties, both theoretical and practical, in tracing their homologies. Before discussing this problem further, the functional role of facial display in primate interaction must be considered.

THE EYES HAVE IT

Adam Kendon has analyzed the facial displays of a kissing couple seated on a park bench with their bodies in contact, who were recorded on 16mm. cine film (1975). This kissing round consists of two film segments, each lasting two minutes. The first two minutes, analyzed in detail, can be subdivided into a complex sequence of events. There are twenty-three episodes of sustained facial expression by the female observable on this two-minute document, utilizing nine different sustained facial configurations or displays. The male's face is much less variable, giving only seven sustained facial configurations, with only minor changes in the conformation of the pursed lips. Moreover, these display episodes take place within periods of involvement in which the participants have their faces turned toward each other. Periods of involvement alternate with periods of disengagement in which the participants are turned fully away from each other. There are ten disengagement events on the record, delimiting periods of involvement, and within each involvement is another series of alternating approach-avoidance turns, with only quarter and half turns of the head. Kendon argues that the record can be further classified into five *phases,* each of which embraces a number of involvements and disengagements, on the basis of the leader of the interaction. Phase one, for example, spans five disengagements and five involvements, all of which are characterized by the female making the first move. In phase two, the male initiates; but the action is immediately terminated by a disengagement of the female. In phase three, the male again initiates, but this phase results in a series of kissing moves, always begun by the male. In phase four, the male again leads but no kissing results, and the male disengages. Phase five is characterized by mutuality in the leading actions, leading to mutual kisses and a mutual termination.

Kendon has implicated the facial expression itself as the controlling factor in these approach and avoidance moves. In phase one, whenever the female turns to the male, she has some version of what Kendon calls the "teeth smile," in which the teeth are exposed. If the male turns to her, they both touch noses and foreheads but do not kiss. If the male turns away, the female leans to him, kisses him on the neck, or leans her cheek against his. Analysis of phase three, in contrast, suggests that the "closed lips smile" by the female cues a kissing approach by the male. In each case, the male waits until the female gives the closed lips smile before beginning his kiss. Moreover, the kissing of phase three ends when the female gives another teeth smile. This analysis of two minutes of social interaction shows that a complex series of approach-avoidance moves is being regulated by mutual monitoring of the face of the other. A similar situation has been discovered in social interactions between children and their care givers.

Developmental psychology has shown that human infants are born with motor control of the facial musculature (Charlesworth and Kreutzer 1973), and they are also using this capacity to modulate the displays of others. By the end of the first month infants are capable of coupling their own tongue protrusion to the tongue protrusion of an adult human (Meltzoff and Moore 1977). Experiments have also shown that both mothers and babies are monitoring each other's facial display.[1] Maurer and Salapatek found that by the second month babies gaze at the eyes of persons, and that the eyes of the mother are preferred, although one-month-old infants fixate the perimeter of the face (1976). Sylvester-Bradley and Trevarthen have shown that mothers' facial expressions and vocalizations closely mirror or complement the baby's own expressions, and when this support is withdrawn, the infant becomes depressed or distressed (Trevarthen 1977). If the mother is asked to hold her face motionless for two minutes, a two-month-old infant may respond with flailing movements and crying. In another experiment described by the Papouseks (1975), involving infants of four months of age, mothers were asked to leave their infants for six intervals of fifteen seconds duration each, behaving during leave-taking as they would at home. The mothers left their babies by engaging in a process of gradual disengagement: maintaining eye contact as they backed away, talking to the child, and so forth. On return, there was a happy reunion with the child, reflected in the infant's smile. If, however, the mother suddenly disappeared, the first reunion was cheerful, as in the normal situation, but repetitions of the sudden leave-taking led to withdrawal by the infant and its turning away. Increased efforts by the mother to renew contact at this stage only intensified the baby's withdrawal. Also using videotape records of mother-infant interaction, Brazelton and colleagues have performed similar experiments with babies from two to twenty weeks of age (1975). The mother is asked to look at the infant with an expressionless face for three minutes. The infant periodically checks the mother's face and then turns away again, and it often becomes listless with a dejected expression. As Brazelton has pointed out, there are analogs to this situation in nature, when sighted children interact with born-blind parents who have expressionless faces. In one reported case of a child with a blind mother, by eight weeks of age the infant developed a mutuality of interaction with the mother that focused on vocal-auditory exchanges. Nonetheless, the pattern of repeated gaze aversions by the infant toward the mother's face was still present at eight weeks. With other people, however, the infant did engage in reciprocal visual interaction.

Brazelton's analysis of these interactions between mothers and two-month-old infants divides the encounters into phases of activity in which smiling and "ooh" faces, vocalizations, play-dialogs, mutual orientation, and disengagements occur in a complex succession.

As these authors note (1975, p. 143):

> The analysis highlights one of the most striking qualities of successful interaction: the mutual cycling of attention and affectivity. The infant demonstrates his participation in this cycling by alternately attending and then withdrawing either partially or completely. The sensitive mother likewise regulates her own behaviors so that her affectivity cycles along with her infant's.

The mutual interaction between display and observer has also been experimentally established in nonhuman primates. Miller devised a "cooperative conditioning" experiment to analyze the physiology of communication between pairs of rhesus monkeys (Miller 1971; Miller et al. 1963). The animals had to discriminate between visual stimuli that had been paired with either punishment or reward, and the responding animal had to pull on one or two levers. A pull on one lever terminated the aversive electric shock stimulus, and a pull on the other lever activated a food pellet dispenser. The animal had to pull on one lever or the other, depending on whether the reward or punishment stimulus was presented. When the animals has mastered this discrimination individually, they were placed in separate rooms in a modified apparatus. One monkey, the "stimulus animal," was given access to the conditioned stimuli, as before, but had no response levers. The other monkey, the "responder animal," had access to the levers but not to the stimuli. However, a video camera was trained on the face of the stimulus animal, and the responder could watch its facial expression via a closed circuit video monitor. All animals but one performed correctly in the aversive tests in the cooperative situation by watching the face of the stimulus animal, but the monkeys largely failed to perceive correctly any reward cues. Analysis of the films showed that the aversive situation was associated with frequent but brief glances toward the stimulus, followed by looking away. On the reward trials, in contrast, the stimulus monkey would fixate his eyes on the conditioned stimulus. This analysis was confirmed by giving the responder monkey access to only the projected film of past performances of the stimulus monkey. In these cases, the responder correctly performed on the aversion trials by watching the film. By counting back from his responses, it was possible to isolate the film segments associated with the bar presses and to analyze them for content. The importance of learning and social experience for judgments of this kind is conveyed by the very different performances given by isolation-reared monkeys when tested in this same experiment. The isolation-reared animals failed to show avoidance to the facial cues of other monkeys, whether fellow isolation-reared monkeys or normals. The monkeys with access to the response lever also performed worse when paired with isolation-reared partners, suggesting that isolates are defective as communicators as well as interpreters. In another experiment, adult male monkeys that had been reared as social isolates were allowed visual contact with unfamiliar stimulus animals and their behavior compared to normal animals in the same situation. It was found that the isolates showed self-threatening, self-biting, self-slapping, and masturbation in these tests, whereas the normals never did (Brandt et al. 1971).

There are also innate components of the perceptual response to facial display, and there would be little reason for innate motor components if this were not the case. Sackett exposed rhesus monkey infants that had been socially isolated from birth, except for brief periods of human hand feeding in the first five to nine days of life, to projected slides (Sackett 1966). The animals were exposed to photos of rhesus monkeys of various ages engaged in various actions, as well as to control slides of neutral topics. Some presentations could be controlled by the infant itself by pulling a lever, whereas other presentations were given by the experimenter. The infants showed a preference for pictures of infant monkeys, compared to adults, and at about two to three months of age, they developed emotional arousal to pictures of threatening monkeys. The responses to the threatening pictures were not simply avoidance but included vocalization and exploratory behavior. These data not only provide evidence for an innate component in the perception of rhesus threat display, but they suggest an innate attractiveness to infant pictures. Lorenz had postulated an innate recognition schema in warm-blooded vertebrates for baby animals, based on the perception of the relatively large head and foreshortened face of babies as compared to adults. Exemplars of the baby schema were postulated to elicit a "cute" response. In this respect, Gardner and Wallach found that American college students judged profiles of babies' heads to be more "babyish" when head-shape was modified beyond the range of head-shape for actual human babies (Gardner and Wallach 1965; Eibl-Eibesfeldt 1975).

In the ethological investigation of innate behavior, innate perceptual and motor components were both postulated by the theory of instinct. The thrust of ethology has been to emphasize that perceptual mechanisms with highly specific eliciting factors coexist with the differentiated world view created by the refinement of the special senses. Some schools of psychology have been so historically conditioned by a philosophical tradition that has equated perception with an objective model of the world as it really exists — a direct descendant of the Age of Reason — that there is still widespread reluctance to grant nonobjective forms of perception an equal place in determining mental content. Emotional perception, if not denied entirely as a distinct process, has been peripheralized to the status of primitive mechanisms, subordinated to a putatively objective world view and preoccupied with vague stirrings in the viscera.

Ethology would argue that this position is without support. In emotional perception, organisms do not attend to all the perceptually distinguishable features of a situation or a stimulus, even features that they are capable of perceiving themselves if tested in other circumstances, but are motivated to attend to certain features at the expense of others. Ethology has demonstrated that some of these forms of selective attention to stimulus features are innate in animals, and one technique of showing this involves the use of models that deviate from the normal. The demonstration that male sticklebacks attack model fish with only a red patch and an eye spot more readily than they attack realistic models is a classic instance of this technique (Eibl-Eibesfeldt 1975; Tinbergen 1953). Such innate releasing mechanisms are widely distributed among birds, and a dramatic example is the preference of plovers for oversized models of eggs, which are so big that the birds fall off when they try to brood them. However, the transformation of Lorenzian instinct into affective cognitions would make such direct elicitations of behavior to a releasing stimulus rare in primates and difficult to demonstrate in naturalistic situations. This is generally the case. Monkeys may give avoidance responses or vocalizations to tape recordings of warning and spacing calls (Waser 1975), but they typically ignore recorded vocalizations when they can see the tape recorder. An apparent exception to this generalization is the tendency of some primates to display to their mirror images. MacLean reported that squirrel monkeys *(Saimiri sciureus)* of the Gothic subspecies or variety gave the penile erection display of their mirror images if kept visually isolated from other monkeys, but the Roman variety of this species did not (MacLean 1964). Also, Gallup reports that individually caged pig-tailed macaques *(Macaca nemestrina)* will display to a mirror but such responding in socially housed animals is infrequent (Gallup 1975; Hall 1962). However, an experiment that placed rhesus monkeys in social isolation for two hours a day, each with the opportunity to open a door to look at either another monkey or its own mirror image, suggests an alternative interpretation to the innate releasing mechanism. Normal monkeys choose another monkey over a mirror image, whereas just the opposite is the case for isolation-reared monkeys. This suggests that mirror display in normal monkeys may be a case of social behavior directed to a less than optimal stimulus due to temporary social deprivation in otherwise normal animals. Gallup argues that the concept of a mirror image may have to be learned in humans, and this interpretation is supported by his study of naive chimpanzees exposed to a mirror. Over a ten-day period, social acts directed to the mirror dramatically decline, whereas self-directed behaviors dramatically increase. Once experienced with mirrors, chimpanzees will use the mirror image to direct behaviors to objects they cannot otherwise see, such as the top of their heads or objects lying at an angle outside of the cage. Using dye marks applied under anesthesia to parts of the body that cannot be seen without a mirror, it is possible to show that chimpanzees develop a mirror image concept, for they touch the dye mark on themselves and not on the mirror. Similar attempts to demonstrate self-recognition in monkeys have failed. These data suggest that mirror display in monkeys is not evidence for an innate releasing mechanism in a classical ethological sense but is rather other-directed behavior by a species that has not acquired the concept of a mirror image.

The lack of consistent overt behavior to species-specific signals by primates has sometimes been taken as evidence that ethological concepts are inapplicable to these species. However, it is far easier to account for the evolutionary continuity of primate social behavior if it is assumed instead that primates lack an invariant connection between innate signal and innate response. An ethological approach would argue that homologs to innate releasing mechanisms are in fact present in the primate brain as modularized components with preferential connections to emotional thought and behavior. These innately salient perceptual components coexist and interact with the fine-grained pattern recognition and conceptual learning associated with the evolution of the special senses. Ethologists have drawn largely upon the behavior of human infants for evidence of innate perceptual mechanisms in man.

Experiments with human infants have shown that smiles are elicited by specific stimulus conditions that represent only a fraction of the situation objectively discernible by adults. Spitz and Wolff found that a mask or face with two eyes, a nose and a forehead was both a necessary and sufficient stimulus for evoking initial smiling in an infant at about two months of age, and a nodding configuration was especially effective (Wolff 1963; Vine 1973; Washburn 1929; Robson 1967). Ahrens found that a pair of dots on an oval or square face cutout was the best elicitor of the first smiles, at about six weeks in his institutionalized infants.[2] Moreover, at this young age, a face cutout with six spots was as good or more effective than one with just two. Although it does not matter if the two-spot configuration is presented horizontally or vertically, a single dot is ineffective. However, as the child matures, the smile becomes more responsive to actual facial configurations, so that by two or three months a human face is the most effective stimulus. Several dozen subsequent studies by many researchers have confirmed that the facial configuration, particularly the eyes, is a potent elicitor of smiles in infants, but other sensory modalities such as voice and touch are also effective, especially in very young infants (one to eight weeks) or when combined with the reciprocal social interaction already discussed. The three-month-old infant is at a peak period for the smiling response, and the response gradually wanes almost completely by the end of the first year. As the smile consolidates, there appears to be an optimal level of social interaction, as both prolonged stimulation and the disappearance of the face can lead to aversion, as indicated by crying and distress movements in the infant. Distortion of *familiar* faces can also lead to crying between three and six months, which suggests that fear and distress are resulting from violated expectations. However, given the centrality of the face schema to the primate emotional system, it is not surprising that distorted faces are potent elicitors of fear in adult chimpanzees (Hebb 1961, p. 243). Moreover, distorted faces are widely used in horror movies, presumably because of their fear-inducing effects on humans in contexts of suspended disbelief. In a rhesus monkey reared on an inanimate surrogate mother, alteration of the dummy's head by the addition of a painted facial configuration produced extreme terror, and the animal consistently rotated the head on the dummy so the face was in back (Harlow and Suomi 1970).

The paired eye-spot configuration is not only salient to human infants but occurs widely in nature as markings on fish, snakes, insects, and turtles (Coss 1970; Hindmarch 1973; Blest 1957; Wickler 1968). Blest tested laboratory-reared birds, who had no opportunity for learning from experienced birds, on *Nymphalis* butterflies, which have eye spots on their wings (1957). The birds reared back when the butterfly exposed its wings, suggesting that the patterns evolved on the insect as part of an antipredator strategy. If the wing spots are removed, the birds will eat the insects. In other experiments, Blest projected the eye-spot pattern next to mealworms that birds were about to eat and successfully scared the birds away. Although there are no experimental data on primate species, the evidence indicates that the paired eye-spot configuration is functionally related to emotional behavior in both humans and birds. However, the period of maximal smiling is also a period of rapid maturation of visual-manual coordination, which argues for the importance of cognitive factors (White and Held 1966). Vine argues that it is not necessary to postulate any special eliciting properties to the paired eye-spot configuration in human infants (1973). Instead, early smiling is a recognition response elicited by successful matching of visual input to an internal model of a familiar object. Since the infant's own eyes are rather poor, it can only resolve the most salient visual stimuli, and the response-contingent action of the care giver gives the infant the necessary input for the abstraction of the eye-spot schema from the flux of the visual world. This cognitive approach is in fact compatible with an ethological approach, since there is no doubt that emotionally salient stimuli in man are cognitive as well as emotional, and there is good evidence that facial recognition, at least, is a neocortically mediated process. Primacy of the eyes and their integration into constructional ability is provided by the drawings of children, where the eyes appear first in pictures of the face, to be supplemented in older children by other features (Gesell 1940, pp. 153–154; Robson 1967).

The best evidence for neocortical processing of facial recognition is provided by patients who have had surgical section of the corpus callosum for epilepsy (Sperry 1974; Levy et al. 1972; Gazzaniga 1970). With these subjects, it is possible to give visual information to a single cerebral hemisphere without the information being transferred to the opposite hemisphere. This unilateral input procedure requires controlled testing conditions, and it is not operative in everyday life. However, in the laboratory, composite visual stimuli can be devised that result in the left hemisphere seeing one picture and the right hemisphere another. By examining which stimulus controls the motor response in the test, it is possible to assess the relative dominance of the two hemispheres for different kinds of input. These experiments show that the right (nonspeech) hemisphere is superior to the left (speech) hemisphere in matching of nonlinguistic shapes and patterns, including faces. The left hemisphere controls when presented with verbal material, whether spoken or written. This distinction between verbal and nonverbal cognitive function is in accord with a wide variety of other data to be discussed later.

Moreover, some evidence suggests that a developmental theory of the facial schema that postulates identity with other kinds of cognitive processing of visual gestalts may be incorrect. After about the age of ten, people lose the ability to make ready identifications of photographs of faces when presented upside down, whereas photographs of objects are easily recognized in altered positions. In fact, the better a normal adult is at recognizing photos of faces, the more he is impaired by inversion of the picture, whereas this relationship is not true of nonfacial material (Yin 1970). Patients with gunshot wounds in the right tempero-parietal area of the right hemisphere do worse on facial recognition than other patients with lesions elsewhere but are *less* disturbed by inversions. Also, a specific syndrome called *facial agnosia* has also been claimed for right hemisphere lesions (Teuber 1975, pp. 462–463). Since facial recognition must be a subtle process, utilizing fine points of distinction, it is necessary to control for the complexity of facial recognition tasks before accepting a distinct mechanism. What these data do prove, however, is that facial recognition is a neocortically mediated process lateralized in the right hemisphere, indicating a cognitive schema of which the eye-spot configuration is likely to be a component. A purely cognitive theory, however, does not address itself to the problem of why a substantial portion of the neocortex should be devoting itself to distinguishing differences between individual humans in the first place, nor does it explain the long mirror-gazing of isolation-reared monkeys and chimpanzees studied by Gallup. There are other ways of organizing social aggregations that do not depend upon individual recognition, and bureaucratic rationalization has moved humans in that direction. Nonetheless, ethology would regard the propensity for individualized discrimination of others as an innate component of the primate biogram. Human beings are innately interesting to other human beings, and this provides the motive power to cognitive discrimination. It would also argue that the two-eye configuration is not only neocorticalized but remains in the adult, as in the infant, as a modularized perceptual component, phylogenetically homologous to similar mechanisms in other animals. These perceptual modules are activated by species-specific signals, but they result in affective arousal by the organism, not necessarily in overt behavior. Rather than ask whether the eye-spot pattern is innately salient, an ethological approach would ask why the first social approaches of a human being should be directed to a stimulus configuration innately associated in other vertebrates with the induction of fear. It is this question that permits a solution to the problem of homologizing the smile.

In attempting to homologize laughter and the smile, Van Hooff draws attention to the morphological similarity of the human smile to the silent bared-teeth face of nonhuman primates on one hand and the similarity of human laughter to the relaxed open-mouth face on the other (1972, p. 238). However, in humans these two expressions converge through a wide range of intermediates. Various subdivisions of the smiling repertory have been made, but one categorization found in several authors under different names is the *simple* smile with only upturned lips and the *broad* smile with the teeth exposed. Although there is a motivational analog among the smiles to the silent bared-teeth display of nonhuman primates, namely the anxious smile known in American anthropological jargon as the "coprophagic grin," the smile that is morphologically most similar to the primate fear grin is the human *broad smile* (Brannigan and Humphries 1972). Oddly enough, observers agree that the broad smile is characteristic of humans at their most happy, socially relaxed, and joyful moments; and it is not motivated by fear. This disassociation could be construed as evidence for a lack of correspondence between the form of displays and the underlying motivation, as critics of ethology have argued, and it is supported by the equivalent disassociation between the induction of fear by the eye-spot pattern in birds and the perceptual fascination for the same pattern in human infants. Van Hooff, considering smiles and laughter, invokes an ethological argument to account for these anomalies. He arranges these two expressions on gradients of friendliness and playfulness, with mouth-opening and vocalization increasing with playfulness and baring of the teeth increasing with friendliness. However, he concludes that the smile, "originally reflecting an attitude of submission . . . has come to represent non-hostility and finally has become emancipated to an expression of social attachment or friendliness which is nonhostility *par excellence*" (1972, pp. 235–236). This argument, a form of motivational shift, neither explains why such emancipation should have come about nor why the smile should have further converged with the relaxed open-mouth face of play. There is another explanation: that the silent bared-teeth face has not changed its motivation in man at all. This argument, seemingly contraintuitive, would maintain that the broad smile is in fact as reflective of fear as its phylogenetic antecedents suggest it to be. The association with social attachment situations is explained by the hypothesis that many "social attachments" are in fact *approach-avoidance conflicts stabilized at a rewarding midpoint.* What has altered in hominids is the reward value of inhibited agonistic consummatory behavior.

FINELY TUNED AGONISM

As Harlow and his colleagues have shown, social isolation of rhesus monkeys for their first six months produces a highly abnormal animal that clutches itself and curls up into a ball in the corner (Harlow and Harlow 1965; Harlow et al. 1965; Mitchell 1970; Arling and Harlow 1967). Both contact play and threat behaviors are depressed to near-zero levels for two months after the animals are introduced to social situations, and this contrasts dramatically to the play and aggression of normal monkeys observed at the same time. A characteristic feature of the six-month isolates was the extent to which they went rigid in social interactions. The isolate would turn away, avert the eyes, and stiffen the whole body. When rhesus are isolated for their whole first year, they exhibit extremely high levels of the bared-tooth face or fear grimace and very high levels of threat but no physical aggression. In fact, one study had to be terminated because the normal monkeys could bite the isolates without fear of reprisal. If tested later in life, these monkeys eventually develop a hyperaggressiveness to complement their hyperfearfulness. A slightly different syndrome is produced by rearing rhesus monkeys in partial social isolation for the first six months and total isolation for the next six months. These animals show increased fear and aggression. They are more aggressive than normal animals, as indicated by the number of social encounters exhibiting physical aggression, but they are also more fearful, as indicated by the number of threats and fear grimaces. Their aggressive behavior often appeared irrational or unnecessarily severe, such as brutal attacks on juveniles or attacks directed to animals much bigger than themselves. In socially isolated monkeys, fights are also of longer duration, of greater severity, and more frequent than in normally reared animals. However, isolation-reared males, in spite of being more aggressive, are more likely to lose dominance encounters with normal males of the same age (Mitchell 1968). It has also been reported that status hierarchies are less stable among isolation-reared monkeys, although this has been questioned. In maternally separated but peer-raised rhesus monkeys, the hierarchies are more viciously maintained but they are not unstable (Mason 1963; Goldfoot 1977). Also, isolation-reared primates are deficient to normals in interpreting the facial expressions of others, as revealed by Miller's (1971) experiments, and isolation-reared females become abnormal mothers who are extremely aggressive and punitive to their own infants. The aggression directed toward infants decreases as the mother gets older, and one explanation is that the females interpret the approaches of their own infants as threats and respond aggressively. Isolation-reared primates are also unusual in the extent to which their instinctive behaviors are self-directed. Harlow has recorded one case in which the monkey appeared to be afraid of its own hand, and self-biting and self-mutilation are common. All of these lines of evidence indicate that isolation-reared primates are abnormal in modulating their levels of fear and aggression to social stimuli, are likely to turn it inward, and are incapable of maintaining fine gradations in their agonistic interactions. They tend to be either cowards or bullies. Gary Mitchell has succintly summarized the results of dozens of such studies (1970, p. 243):

> Isolation-reared primates change as they mature from rocking, digit-sucking, grimacing, self-clutching recluses to pacing, socially aggressive, self-threatening, masturbating, self-mutilating menaces who often make bizarre movements.

The effects of social isolation suggest a deficiency of the agonistic system in modulating facial display, biting, and locomotor behavior. Agonistic behaviors and the situations that trigger them appear to have very strong innate components, whereas the fine tuning of the agonistic system is a function of social experience. Kruijt (1964) found a similar effect among jungle fowl (*Gallus*). Isolation did not interfere with the development of the motor patterns of agonism but with their frequency, modulation, and integration. It will be recalled that mammals with all brain tissue above the midbrain removed, except for an insular hypothalamus, are capable of motor components of agonistic behavior, although they lack a normal time course and sensory guidance.[3] Also, lesions of limbic system forebrain mechanisms in mammals alter the expression of agonistic behaviors without abolishing the behaviors themselves. Some of these experiments show very close parallels to the results of isolation rearing.

Lesions of limbic forebrain areas in mammals produce marked disturbances of the frequency, threshold, and situational appropriateness of instinctive behaviors, as well as effects on motivation, activity level, and behavioral inhibition (Kaada 1967; Kling and Mass 1974; Kling and Steklis 1976; Myers 1972; Pribram 1961, 1969, in press; Rosvold et al. 1954; Schreiner and Kling 1953; Slotnick 1967; Slotnick and Nigrosh 1975). Destruction of one component of the limbic system, the amygdala, usually produces a profound taming of otherwise wild animals, although in some cases, the opposite effect, increased ferocity, has been obtained. Schreiner and Kling provide dramatic evidence of hypersexuality and tameness after amygdala lesions in a variety of species, and it is known that this structure can modulate the time course and intensity of aggression produced by electrostimulation of the hypothalamus. In rodents, lesions of the septum, a limbic area with close connections to the hippocampus, also produce hyperaggressiveness or hyperfearfulness, depending in part on the species of animal. The neocortical components of the limbic system — the prefrontal lobe and temporal pole — have been much studied in primates, and lesions in these structures produce monkeys with a flattening of affect, aimless hyperactivity, and disinterest in the social environment, quantifiable as a reduction in the frequency of facial expression and vocalization. Amygdala lesions typically produce a drop in social rank in monkeys, as would be expected with taming, but in some cases the animals retain their rank or even rise, depending in part upon the social situation at the time of their reentry and on the sex of the animal. In complex social environments, amygdalectomized monkeys withdraw from active participation and become social peripherals. All of these experiments are consistent with the interpretation that limbic forebrain mechanisms provide control of the frequency, threshold, duration, and perhaps targets of the behaviors involved in fear and aggression.

Pribram has interpreted these lesions as altering set-points in the homeostatic regulation of emotional behavior (1967, 1969). The importance of this approach is that it postulates a tunable homeostat, adjustable by alterations in the set-point, and not a fixed level of optimal emotionality. However, Pribram's initial formulation of this model of affect emphasized the homeostatic regulation of sensory input. Although Pribram has justly criticized theories of emotion that rely on a visceral brain and has replaced it with a model that emphasizes the regulation and control of external information, this model still does not take cognizance of the ethologist's fundamental point of departure: *that emotions are intrinsically social phenomena.* The social deprivation, which is so incapacitating in the development of social behavior, does not produce a comparable deficit in intellect. Isolation-reared monkeys, when tested on sensory discrimination problems, perform as well as normals provided that care is taken to habituate them to the testing situation (Harlow et al. 1971). If emotions are systems for intelligent social responding, one would expect them to be disrupted by restricted social experience in a way that mechanical forms of intelligence are not.

Although disorders of physiological homeostasis have been observed after limbic lesions, the emotional component of the limbic forebrain is not primarily oriented to the maintenance of visceral stability. Rather, the tunable homeostats of the emotional mechanisms are involved in the creation of stable response states that counterbalance the instinctive needs of the animal with the contingencies of the external social environment. A good example of such stable modulation is provided by the analysis of defensive threat, which is an emotional state that can be maintained for relatively long periods of time. Blurton-Jones, in his analysis of the conflict behavior of the passerine bird *Parus major*, has shown that threat display can occur when attack is blocked by any agent, whether a physical barrier in the external world or by the simultaneous induction of fear (1968). The emotional display thus represents a state of balance between countervailing impulses that incorporate external and internal information simultaneously.

Threat display is a stable midpoint on an approach–avoidance gradient that is widespread among vertebrates; but in normal, socialized primates, fine tuning of the agonistic system provides a number of such stable midpoints, from low intensity threat to savage attack and from panic-stricken escape to a brief aversion of the eyes. Fine tuning allows these variants to be highly contextualized to specific classes of environmental events, and it permits the agonistic system to react to small agonistic actions of others with levels of defense, attack, or withdrawal commensurate with past experience. The result of this fine tuning is a highly differentiated agonistic response repertory directed to a highly differentiated social world, in which most agonistic transactions are conducted with intention movements and threats and never reach high intensity consummatory levels. This spectrum of possible agonistic responses with their associated contextual cues will be referred to subsequently as the *agonistic tuner*. In humans, the emotional reactivity system is closely related to the conceptual system itself. Alterations in purely conventionalized social relations can lead to emotional reactions. The sociological school of ethnomethodology has conducted small "field experiments" that violate conventional rules of interaction, as in the following: (Garfinkel 1967, p. 43):

CASE 3

"On Friday night my husband and I were watching television. My husband remarked that he was tired. I asked, 'How are you tired? Physically, mentally, or just bored?' "

(S) I don't know, I guess physically, mainly.
(E) You mean that your muscles ache or your bones?
(S) I guess so. Don't be so technical. *(After more watching)*
(S) All these old movies have the same kind of old iron bedstead in them.

(E) What do you mean? Do you mean all old movies, or some of them, or just the ones you have seen?

(S) What's the matter with you? You know what I mean.

(E) I wish you would be more specific.

(S) You know what I mean! Drop dead!

The effect of this fine tuning is to convert agonistic behavior from a predominantly social dispersal mechanism to an affiliative mechanism involved in the social bond.

THE AGONISTIC SIDE OF LOVE

As a number of authors have observed, agonism is not only affected by the social bond; it is intimately involved with the maintenance and perhaps the establishment of the social bond (Kagan 1978; Bowlby 1958; Hinde 1961). Fear appears to be critical in the formation of the initial care giver-child attachment relationship by providing the cutoff point for the development of a bond or reinforcing the bond once it has formed. In primates, a generalized disturbance reaction to noxious situations is present at birth, such as the crying reaction in the human infant. The diffuse distress reaction is, ethologically speaking, a request for intervention on the part of the care giver. This interpretation is compatible with the finding that wild-caught chimpanzee infants that have had social experience give more distress vocalizations than isolation-reared animals (Randolph and Mason 1969). The emotion of fear, in contrast, is a motivational and response system related to a certain level of competence on the part of the animal. The manifestations of fear in rhesus monkeys increase with age; the diffuse distress behaviors decrease with age but increase with frustration (Bernstein and Mason 1970). When placed in frustrating situations, such as food removed beyond its reach, the immature monkey may respond with self-directed and distressed behaviors that are a regression to the response repertoire of the younger infant. It is in accord with this "competence" view of emotion that the diffuse distress reactions are present at birth in primates, as well as in carnivores and some birds, but fear reactions, as indicated by the species-specific signals, do not appear until later in infancy (Mason and Berkson 1962; Bronson 1968; Hinde 1961; Hansen 1966; Kagan 1978; Salzen 1967). Studies show that fear only develops in human infants during the second half of the first year, in rhesus monkeys not before thirty days, and in dogs not before two weeks. Many authors have related the ontogeny of fear to the development of social attachments and to the differentiation of the perceptual world into novel and familiar. The first fear responses are often given to strangers, and their appearance coincides with the differentiation of distinct attachment relationships to particular individuals. This interpretation is also supported by studies of imprinting in birds, which show that the end of the critical period for imprinting coincides with the onset of fear. Once fear has developed, strange stimuli, which formerly had elicited approach or interest, can elicit withdrawal or fear. However, it is well known that the fear or withdrawal responses are reduced by the presence of attachment figures. The attachment figure provides security for play and exploration, and, conversely, isolation-reared primates are extremely fearful in strange situations.

Sackett studied the effects of social deprivation on exploration and responsiveness to novelty (1972).[4] Rhesus monkeys, who had been reared in different situations during their first year, ranging from a wild troop to twelve months of social isolation, were exposed to novel environments and different kinds of visual stimuli as young adults. Preference for complex stimulus patterns increased with the level of social complexity of the rearing condition, and deprived monkeys showed less activity in the novel environment. There is also a relationship between limbic system lesions and the reactivity to novelty, which has been described by Pribram, who incorporated this phenomenon into his model of emotion. He theorized that the core brain systems that maintained physiological homeostasis also provided a hyperstable "ground" that allowed the minor perturbations of novel "figures" to be recognized. The ethological evidence gives partial support to this concept, but it suggests that the background against which levels of novelty are measured is not physiological homeostasis but the tonic inhibition of fear reactions, induced by vestibular and tactile biasing of the limbic system in the course of social experience (Mason 1973; Melzack 1965; Hofer 1978). In a more recent formulation, Pribram has argued that the assessment of novelty is in fact specifically involved in the central pathways of diffuse temperature and pain stimuli that terminate in the amygdala (Pribram in press).

The rearing of primates in social isolation has demonstrated the primacy of tactile stimulation in the response system of the nonhuman primate infant. Isolated infants prefer a surrogate mother covered with cloth to a less "tactile" surrogate that dispenses milk, and holding and touching are the usual techniques for calming human infants. Moreover, holding infant chimpanzees markedly reduces the stress reaction, and the stereotyped rocking behavior of isolation-reared primates can be prevented by rearing them on mechanically mobile mother surrogates. Holding and clasping is also common in the comforting behavior of primates in naturalistic situations, and it has also been demonstrated that sensation from the skin and muscles can modulate the activity of the reticular formation (Gastaut and Bert 1961; Pompeiano and Swett 1962a, 1962b). Kaufman and Rosenblum, following the clinical psychology tradition, have referred to the primate attachment figure as a "psychotaxic object," and this terminology calls attention to the fact that the care giver functions as a psychological entity with power of its own (1967). Since the recognition of attachment figures requires perceptual learning, and since facial recognition is in part a neocortical process, the attachment figure is no simple stimulus but a perceptual–cognitive entity that can exert a tonic (or sustained) inhibitory effect on the reactivity of the agonistic system.

An interrelationship between agonism, attachment, and exploration is not surprising when viewed from the perspective of approach–avoidance modulation (Schneirla 1959). The locomotor behaviors associated with fear and aggression differ in their polarity. Fear moves the animal away from the source of stimulation, but the consummatory behavior of aggression requires approach toward the source of unpleasantness. Consequently, any modulation of fear and aggression entails a modulation of approach–avoidance behavior. Attachment figures, by providing tonic inhibition of agonism, would also affect the dynamics of exploration. The precise effect, however, would also be influenced by the relative preponderance of fear and aggression in the course of maturation. When fear is the predominant agonistic motivation, the reduction of inhibition on the agonistic system, caused by an increase in distance between the partners, will provoke locomotion toward the familiar. As aggression matures, a similar reduced inhibition provokes locomotion toward potentially threatening sources of novelty. Although this is the general trend of development in all species of primates, the characteristics of agonistic behavior and their relationship to attachment show wide variation among primate species, and these interactions must be studied on a case-by-case basis because other factors are also involved. Also, approach–avoidance behavior is the final common path for a number of central systems, as the analysis of gaze direction indicates (table 3.1) (Argyle and Cook 1976). In addition, the categorization of stimuli into novel and familiar requires the full cognitive resources of the animal and is not reducible to the regulation of arousal. Monkeys have been taught novelty learning sets, in which they must always choose the stimulus object that they have never seen before, and learning set ability is phylogenetically progressive in mammals.[5] Also, the sensory discrimination operations, which are logically prior to correct performances on such tests, are also neocortical functions in primates, as they are profoundly disrupted by lesions of association cortex for the specific modalities of sensation.

This ethological approach, which construes the attachment figure as a perceptually monitored conceptual entity that exerts tonic inhibition on the reactivity of the agonistic system, is in accord with both ethological theory and primate attachment behavior. As classical ethology maintained, low levels of action-specific energy are manifested by part behaviors and intention movements taken from a more complex instinctive act. Similarly, agonistic activation often produces simultaneous fear and aggression or alternations between the two states that are revealed by both ambivalent gestures and approach–avoidance oscillations. Consequently, the criteria of

TABLE 3.1: *Determinants of Gaze Direction*

A may look a lot at B under the following conditions:

> they are placed far apart
> they are discussing impersonal or easy topics
> there is nothing else to look at
> A is interested in B, and in B's reactions
> A likes B
> A loves B
> A is of lower status than B
> A is trying to dominate or influence B
> A is an Arab, Latin American, and so on
> A is an extrovert
> A is affiliative (and female in a cooperative situation)
> A is low in affiliation (and female in a competitive situation)
> A is dependent (and B has been unresponsive)

A may look very little at B under the following conditions:

> they are placed close together
> they are discussing intimate or difficult topics
> there are relevant objects or an interesting background to look at
> A is not interested in B or in B's reactions
> A dislikes B
> A is of higher status than B
> A is from one of certain American Indian tribes
> A is an introvert
> A is autistic, schizophrenic, or depressive
> A is affiliative (and female in a competitive situation)
> A is low in affiliation (and female in a cooperative situation).

SOURCE: from Michael Argyle and Mark Cook, *Gaze and Mutual Gaze.* Cambridge University Press (1976), p. 176. With permission.

finely tuned agonism—grading of agonistic response with part behaviors substituting for full-blown aggression and fear—are precisely the behavioral phenomena that would be predicted from tonic inhibition of agonistic intensity. Moreover, this theoretical approach suggests that the outer boundaries of attachment relationships should be demarcated by "real" agonistic behavior: by high intensity fear or aggression. The further we move from attachment figures, either spatially or figuratively, the less the inhibition on agonistic reactivity and the more likely that avoidance or fighting will occur (Marler 1976; Hrdy 1977). Although this corollary is also generally true of primate behavior, it is confounded by an additional property of attachment relationships not yet discussed: in primates, agonism is not only a function of attachment "distance" but is also widely used to maintain boundaries to attachment relationships that might otherwise be breached or to create boundaries where none existed before (Bernstein and Gordon 1974).

Pigtail macaque mothers will aggress against other females who approach their newborn babies, and most species attacks on infants are responded to as if they were attacks on the care giver. Primate infants often throw temper tantrums at the birth of a sibling, and in some cases offspring will harass males attempting to copulate with their mothers. The attempted withdrawal of the attachment figure can also provoke aggression; and hamadryas males, who bite their females on the neck when they attempt to stray, provide a good example. In all of these cases, aggression is in the service of attachment; but it can also work toward the opposite effect. The same pigtail mothers who jealously guard their new infants will actively rebuff them with cuffs and mild biting when the babies are between five and eight months old, and intergroup agonistic encounters are a widespread feature of primate social organization (Rosenblum 1971). In these cases, aggression prevents the formation of certain kinds of social bonds or severs bonds that are already functioning. Primate attachment does not only have inhibitory effects on agonism; but as Bernstein and Gordon have pointed out, aggression is systematically used in the maintenance of attachment relationships.

The role of aggression in attachment varies among primate species, and there are well-documented psychological differences, probably in part genetic, that affect the intensity of social bonds and help create some of the variations seen by field primatologists (Rosenblum 1971; Sackett et al. 1976). In the laboratory, *Macaca nemestrina*, for example, forms tightly knit mother-infant dyads, with high maternal protection and high levels of contact behavior, followed by high levels of maternal punitiveness and severe separation stress in the infant. In contrast, the closely related *Macaca radiata* differs on all of these measures. Similar differences in attachment strategies have been reported for primate species in the field. Although behavioral observations do indicate a systematic connection between agonism and attachment, they also show that the relationship is a complex one, with a number of levels interacting simultaneously.

This "tonic inhibition" approach to attachment also allows for the possibility that other kinds of experience could substitute for developmental deficiencies. Suomi has shown that even the dramatic behavioral deficits created in rhesus monkeys by total social isolation during the first six months can be reversed by housing the animals with three-month-old rhesus females (Suomi and Harlow 1972). These infants, who are themselves immature and inexperienced, do not aggress against the newly emerged isolates but provide a source of reciprocal stimulation and tactile contact. This allows a social familiar to be differentiated incrementally and provides an inhibitory ground for the subsequent assessment of novelty and the development of finely tuned agonistic response. Similarly, tonic inhibition also explains certain violent acts that have provided support for the regressive interpretation of human instincts. Unilateral withdrawal from intense attachment relationships should disinhibit the agonistic system and give scope to the full intensity expression of consummatory aggression, perhaps even with a postinhibitory rebound to higher than normal levels. Such irrational acts are only irrational when viewed in terms of the benefits to the actor but not when examined in the context of a theory of primate aggression.

PRECONSUMMATORY JOY

In addition to social interactions involving actual agonistic behavior in either the intense or finely tuned forms, there are also distinct affiliative systems in primates that appear to be phylogenetically derived from the agonistic system. These agonistic derivatives can also be accommodated within the concept of tuning mechanisms, but they entail innate set points for "inhibited" aggression and fear that are distinct from the fear-induced inhibition of actual agonistic behavior. The affiliative instincts that derive from agonism are characterized by the stabilization of attack or fear behavior without the aversive affective consequences. These affiliative derivatives of agonism are usually referred to as social play.[6]

As many ethologists have noted, the play behavior of mammals incorporates the behavior patterns of attack and fear, but the participants do not usually get hurt. Moreover, as Altmann has observed, there often appears to be a self-handicapping involved, in which individuals moderate the strength of their behavior when playing with smaller partners (1961, 1962; Bekoff 1972). The participants appear to enjoy the activity: they return to it again and again, and they do not give the facial expressions and vocalizations that occur in actual agonistic behavior. Furthermore, the participants frequently alternate roles, such as chaser and victim, whereas real agonistic behavior has an asymmetrical termination. The facial expressions and vocalizations that do occur are largely specific to the play context, and if they occur elsewhere it is in relaxed or exploratory contexts. If actual agonistic signals occur in the course of play behavior, the usual effect is to disrupt the activity or bring intervention by adults. In monkeys, too, the play behavior develops before the maturation of social aggression. Also, in play, pauses in the behavior are likely to be filled by solitary or parallel forms of exploration and locomotion, which is not the case with actual aggression. All of these properties have led students of behavior to classify play as a distinct category. At the same time, the similarity of the behavioral components of play to agonistic behavior has suggested an affinity to the latter. In nonhuman primates, the major activities of social play are chasing, inhibited biting, jumping on the other, cuffing, pulling and rolling on the ground while grappling — all of which occur in aggression. The facial expression of play also looks like an intention movement of biting. Moreover, Van Hooff's analysis of chimpanzee activity concluded that play was an affiliative complex linked to aggression through the motor pattern of hitting. Also, in polecats, play and aggression are directly related, in that social play experience influences the form of the subsequent biting attack directed to prey. However, the consummatory behavior of aggression is inhibited in play. In a cinematic study of polecats *(Mustela putorius),* Poole found that the major difference between play and agonistic behavior was the frequency of inhibited biting. In play, biting occurs only 2 percent of the time, compared to 40 percent in aggression (1978).

The infliction of pain on another individual is the consummatory behavior appropriate to aggressive motivation. For humans at least, this is a more appropriate way of characterizing aggressive motivation, because humans employ their intelligence to devise new and interesting ways to inflict pain, which cannot be reduced to the innate consummatory behaviors associated with aggression. This suggests that the goals of innate consummatory behaviors in humans are represented conceptually as if they were external actions and then integrated with the instrumental modality. Nonetheless, biting and hitting are innately linked to aggression in all primates, including man, and these innate consummatory behaviors have come to function in an innate system of ritualized aggression. If the behavior appropriate to biting attack is inhibited by the jaws failing to close, then social encounters can be prolonged at an intimate distance by stabilizing this *preconsummatory phase.* The ritualized combat of male hamadryas baboons uses this principle, and the males spar by directing their mouths to each other's necks (Kummer 1968). Film analysis shows that the bites do not in fact reach the consummatory stage. However, this behavior is apparently agonistically motivated, even though injuries are reduced to a minimum. In play behavior, there is a similar inhibition of biting, but the behavior appears to be a distinct modality with enjoyable motivational properties. The concept of agonistic fine tuning suggests an explanation. If the thresholds of the biting response are raised by limbic system control of the peripheral motor system, then play can be easily explained as a specific set point on the agonistic tuner in which the behavior stabilizes at the preconsummatory phase of agonistic behavior. A possible mechanism for such an interpretation is provided by Flynn's studies of the neural control of biting attack in the cat (1967).

In cats, rats, and probably in other mammals that are potentially carnivorous, both affective attack and quiet-biting attack can be elicited by electrostimulation of the brain. Affective attack shows all the indications of emotional arousal and display, whereas quiet-biting attack is a cool, goal-directed behavior. In cats, biting attack requires control of the theshold of reflexive biting, and this in turn is mediated by the trigeminal nerve. The trigeminal or fifth cranial nerve is composed of both sensory and motor branches. It carries tactile and pain information from the teeth and skin of the upper and lower jaws and controls the contractions of the associated muscles. Flynn found that if an adult cat is stimulated in the area of the hypothalamus, which produces biting attack, then a tactile probe to the face adjacent to the upper lip, in the region between the nose and the vibrissae, elicits head movements that bring the lips into contact with the object. Tactile stimulation to the upper or lower lip causes opening of the mouth and a strong bite. This reflex can be abolished in adult cats by severing the sensory branches of the trigeminal nerve. Once this nerve is cut, not even stimulation of the biting attack area can elicit more than a single bite in some cats, and in others biting is abolished entirely. Furthermore, in cats with an intact trigeminal, stimulation of the hypothalamus increases the size of the field from which the mouth-opening response can be elicited. At low levels of stimulation, the probe must be in the region of the canine teeth, whereas at high levels, biting can be elicited from the corners of the mouth. Also, electrostimulation of the motor nucleus of the trigeminal nerve causes the jaw to close, but the jaw closes more fully if this stimulus is paired with stimulation of the hypothalamus. A role for the trigeminal system in agonistic response is also suggested by the phenomenon of trigeminal neuralgia, an intense form of central pain subjectively localized to the sensory field of this nerve. The hypothalamic mechanism is itself controlled by limbic system structures of the forebrain. Concurrent stimulation in the amygdala and hippocampus can facilitate or suppress biting attack elicited by hypothalamic stimulation. There is observational evidence that the consummatory behavior of infantile hunger, namely sucking, is also mediated through central modulation of a trigeminal reflex (Prechtl 1958). In human neonates, as well as in other mammals, there is a directed head-turning reflex, which is elicited by tactile stimulation to the corner of the mouth. The head turns toward the source of the stimulation, and the infant's head can be moved about by relocating the tactile stimulus probe. Once contact is made with the lower lip, the sucking reflex comes into play. The threshold of this response is lowered by hunger, and its reflexive status is supported by the fact that it is found in humans born without a forebrain. Prechtl's map of the field from which this rooting reflex can be elicited corresponds to Flynn's map for the biting reflex in cats. These studies provide evidence for limbic and hypothalamic control of threshold levels for reflex systems involving directed head turning, tactile stimulation, and oral consummatory behavior. Further corroboration is provided by Leyhausen's investigations of the development of prey killing in cats.

Kittens whose mothers do not bring them live prey between the sixth and twentieth week of development normally do not kill prey, but they still retain the ability to do it (Leyhausen 1973). Experiments show that the threshold of excitation is higher in adult animals, so that it takes considerably more stimulation to bring the adult to bite forcefully enough to kill its prey; but once this threshold is crossed, it performs the action normally. In kittens, the mutual interaction among siblings, mother, and the prey animal appear to create enough excitement for the kitten to bear down hard with its teeth, and one factor apparently preventing the killing bite is the inhibition produced by the fear of the prey. In a study of prey catching in the European polecat, Eibl-Eibesfeldt found that these animals must learn the orientation of the killing bite, and they also learn the appropriate prey objects (1975). They chase small animals that run away, and they bite them if they catch them. However, if polecats are raised with rats they are inhibited from biting them. Similar findings have been reported with cats, who will treat rats as social companions if they are reared together, and one cat studied by Leyhausen subsequently became a companion to a prey rat who had succeeded in hiding under the cat's sleeping box. The rat would emerge to nibble the cat's heels and duck back under cover. Eventually the rat came to occupy the sleeping box itself and shared the cat's food. This same cat continued to kill other rats, however, and it also killed its former companion when the latter was reintroduced after a three-month absence.

Both neurological and developmental evidence indicate that changes in agonistic behavior can be brought about by limbic system control of the thresholds of innate consummatory behavior. The behavioral similarity of play to aggression is in accord with the view that play is a late phylogenetic emergent, requiring the differentiation of a new, innately determined set point on the agonistic tuner. In primates, the inhibited bite has been further modified into a distinct facial display, the play-face. This interpretation is supported by the report that the play-face is more common in contact play than in approach-avoidance play, as would be expected from this derivation (Chevalier-Skolnikoff 1974; Bekoff 1977). However, social play differs from the other points on the agonistic tuner by being a true affiliative behavior, which is actively sought and apparently does not share the aversive aspect of fear and aggression. In this respect, social play exhibits the same paradox that was already encountered in discussion of the human smile: a form of agonistic behavior appears to have altered its underlying motivation to joy. It was precisely this motivational anomaly that led Van Hooff to invoke motivation shift as the explanation to the smile. However, there is an alternative to the classical ethological interpretation. The anomalies can be resolved by hypothesizing that the *consummatory* behavior of *playful* agonism is the *pre*consummatory behavior of *real* agonism. What has shifted is not the motivational system but the threshold of the consummatory act. Affiliative agonism, by definition, is agonistic behavior whose consummatory act is set at high threshold. Since all consummatory acts are rewarding, the execution of high-threshold biting or high-threshold fear is rewarding, too. However, where the reward value of real aggression requires the closing of the jaws, in the play state the same reward value can be obtained with the consummatory act set at such a high threshold that in fact it never occurs in complete form. Thresholds that are so high that the reward value can occur without the complete consummatory act can be called *preconsummatory reward states* or *hyperthresholds*. Some support for the concept of aggression with a positive reward value is indicated by intracranial self-stimulation in rats (Panskepp 1971*a*, 1971*b*, 1971*c;* Roberts and Kiess 1964). If electrodes are implanted in both the affective attack and quiet attack brain sites in rats, and if the animals are given access to levers that can turn the electric current off and on, the rats will turn *off* the electrodes that elicit affective attack and turn *on* the electrodes that elicit quiet attack. There are also self-reinforcement areas in the brain that elicit continuous lever-pressing to turn on brain stimulation, and the drugs that affect the rates of intracranial self-stimulation can also affect the elicitation of aggressive behavior.

It thus becomes clear why the human smile is morphologically similar to the fear-grin of nonhuman primates, why the smile is similar to laughter at low intensities, and why the human infant's social response system should be directed to a stimulus that is fear inducing in nonmammalian vertebrates. The simple or low intensity smile of humans is the expressive manifestation of the midpoint between two social affiliative systems: hyperthreshold inhibited escape, manifested by the fear-grin, and hyperthreshold inhibited biting, manifested by the play-face. In humans, as in chimpanzees, the fear-grin becomes an affiliative expression that evolves into the wide-smile of humans. The primate play-face, associated with inhibited biting, evolves into the laugh in man, characterized by both the open mouth and staccato vocalizations of the homologous chimpanzee form. These affiliative systems, by definition, are agonistic systems whose consummatory behavior is set at preconsummatory reward level. However, the contextual situations in which laughter occurs in humans suggest that laughter is not simply attack stabilized at the preconsummatory phase but a compound of both attack and fear. Although laughter can be evoked by the misfortunes of others, either real or simulated, as in slapstick comedy, suggesting the consummatory behavior of aggression, it is also evoked in situations of surprise, incongruity, relief, and failed prediction, which are more compatible with fear. The situations that provoke laughter are not only those that provoke attack but those that could potentially provoke agonistic behavior of any kind. This suggests that laughter is an innately differentiated bandwidth on the agonistic tuner within which a spectrum of fearful to aggressive laughing can occur. In other words, laughter is also a tuner that is nested within the agonistic tuner itself. Such modularized approach-avoidance gradients have a precedent in ethology. Fischer found that the "triumph ceremony" of the graylag goose *(Anser anser)* is the product of two conflicted behavioral tendencies: a tendency to "roll" and a tendency to "cackle" (Fischer 1965). However, rolling is itself the product of a conflict between the tendency to escape and the tendency to remain with the partner. In a similar fashion, laughter and play can interact even though each, considered separately, is an approach-avoidance modulator in its own right.

The morphological similarity between the human smile and the nonhuman primate fear-grimace, the rapt fascination of human infants with the eye-spot configuration, and the behavioral similarity between play and aggression are not fortuitous. All these phenomena occur because the social affiliative systems of humans are phylogenetic derivatives of the agonistic system, made possible by forebrain limbic system mechanisms that modulate the thresholds for fixed-action patterns in the course of social interaction. Social attachments are themselves mediated through dynamically controlled set points on approach-avoidance gradients, and the smile and laughter of humans are hyperthreshold derivatives of fear and attack that maintain the interactive distance rewarding to both partners. As revealed by Kendon's analysis of the kissing round, the simple smile functions as a "go" signal to the partner for further interaction. As the threshold of aversion is reached, the simple smile intensifies until it approximates the fear grimace of nonhuman primates. This broad smile functions as a "stop" signal that causes the partner to halt if the interaction is to be maintained. The human broad smile retains its semantic equivalence to the fear-grimace within the mutual social context: its *affect* may be joy, but its *effect* is still the message of inhibited escape: "Come no closer."

PRENATAL BIASING

The affiliative derivatives of agonistic behavior have important implications for primate sociology because they provide one of the best documented examples of sex differences in nonsexual behavior. It was observed quite early in studies of primate social development that male rhesus monkeys initiate more social play interactions than do females, and subsequent experiments have shown that social play is responsive to the prenatal affects of male hormones (Rosenblum 1974; Sackett 1974; Goldfoot 1977; Money and Ehrhardt 1972). If genetically female rhesus monkeys, that is, those with two X chromosomes, are exposed to testosterone while still *in utero,* the external genitalia are masculinized, with penis and scrotum but no testes, and the monkey exhibits behavioral characteristics more typical of genetic males after birth. Mounting behavior, which develops during the first year in normal rhesus monkeys, occurs at male frequencies, as do rough-and-tumble play, aggression, and threat. Studies of genetic female humans who are prenatally androgenized are consistent with this finding, as these individuals show a statistically significant tendency to engage in more socially male behaviors, characterized as the tomboy role, even though they are raised as girls. However, there is experimental evidence that female monkeys may express masculine behavior in the absence of males. If young female rhesus are reared in all-female groups, they not only display more threats than heterosexually reared females but more of the animals will display the male foot-clasp mount, which is the most sexually dimorphic behavior pattern in young rhesus. It is consistent with the hormonal hypothesis, however, that even in the all-female groups, the frequencies are still well below male levels. In the converse experiment, all-male groups of young rhesus show very high levels of presenting behavior. Both mounting and presenting behaviors are not simply sexual in macaque monkeys but are related to social rank, with high-ranking individuals generally doing more mounting and subordinates more presenting. Consequently, the study of sexually dimorphic behaviors in macaques is interrelated with the study of aggression and fear. Further support for this interrelationship is provided by the literature on social deprivation.

A highly consistent finding of deprivation studies is that male primates are more severely impaired than are females, and the most dramatic and long-lasting behavioral disability is the failure to develop the full foot-clasp mount pattern that characterizes sexual mounting in adult male rhesus monkeys. This pattern usually develops in peer play during the first year of life, even though the males do not become capable of breeding until about their fourth year. Complete social isolation during the first year produces adult males who are incapable of an adequate sexual mount, in addition to their profound emotional disabilities. However, even in peer-reared monkeys, without mothers, foot-clasp mounting only occurs in about a third of the monkeys and it makes its appearance one to two years later than normal. This is in spite of the fact that peer rearing does not produce the marked emotional abnormalities of the isolation-reared, and the animals are capable of social play. The deficit is not due to an inability to perform the behavior pattern: the foot-clasp mount appears in infant monkeys who have never seen the behavior performed by older males, and socially isolated monkeys will give foot-clasp mounts to cloth dummies, even though they do not mount other monkeys. Goldfoot has pointed out that peer-raised monkeys show more intense aggression than socially reared animals of the same age, and once again, the evidence points toward inhibition of the agonistic system by care givers. The adult not only provides the security that allows low levels of agonistic behavior among peers, but it actually intervenes in the agonistic encounters of the young. When this inhibitory effect is absent, there is a disruption of the sexual development of male monkeys. It is the interaction between the peer affectional and the maternal affectional systems that is crucial to this development, as monkeys raised with their mothers but only given access to peers for a half-hour a day are also deficient mounters.

There are other lines of evidence that also indicate sex effects on the reactivity of the agonistic system. In the self-directed aggression that occurs after isolation-rearing, males exhibit about twice as much of this behavior as do females. Also, normal male rhesus monkeys in captivity have been shown to be more willing to approach novel objects and to explore novel environments than normal females. These differences are quantitative and not large, but they suggest that males are ordinarily less fearful in the presence of novelty. However, socially isolated males are much more reluctant to explore than either normal females or socially isolated females. These data on exploration, novelty, self-directed aggression, and sexual mounting all point toward an innate sex difference in the reactivity of the agonistic system, with the males being more severely affected by postnatal experience. The greater reactivity of young males to traumatic intervention is also indicated by the fact that lesions of orbitofrontal neocortex impair delayed response and discrimination reversal tests in male rhesus monkeys at two and a half months of age, whereas comparable effects are not found in females until fifteen to eighteen months (Goldman et al. 1974). This is in spite of the fact that there is no sex difference in the performance of normal monkeys on these tests.

Rosenblum, who observed that the same stimuli that elicit approach in male macaques can elicit withdrawal in females (1974), suggests that female monkeys have a lower threshold of fear than do males. However, the social deprivation studies show that the male thresholds are set to a large extent by social experience, such that isolated males may have *lower* thresholds of fear than normal females; and Goldfoot's experiments on same-sex rearing groups also show that female fear thresholds may be lower in the presence of males. Moreover, socialization differences have also been noted, with rhesus mothers playing more with males and restraining more with females. Since refinement of experimental technique usually reveals subtle social effects that were missed by more global analyses, it is likely that future primatological research will reveal further contextual constraints on the reactivity of the agonistic system. The concept of differential fear thresholds is an important one to primate sociology, but rather than absolute threshold levels, the concept of a tunable threshold, set in part by emotional releasing signals in the course of social interaction, would have much more explanatory power when applied to primate behavior. Innate differences are not precluded and can still affect the magnitude of the reactivity to social experience, as the sex differences in monkeys suggest. Rosenblum's concept of a lower fear threshold in females is likely to be true of females in the presence of normally socialized males, and there are innate components to this complex social effect. It is important to note that more fearful females would often create the appearance of relatively more aggressive males, whether or not there are any differences in aggressiveness between the sexes, because situations that provoked males to move forward would cause females to retreat. Among themselves, however, or in familiar situations, females might be expected to be just as aggressive as males are supposed to be. In a four-year study of activity levels in pigtail macaques *(Macaca nemestrina)*, Bernstein found that females were in fact both *more* aggressive and *more* submissive than males, indicating higher levels of agonism generally (1970). However, male aggression was more likely than female aggression to lead to injury, especially from biting. In perhaps a related finding, Hausfater (1975) showed that in savanna baboons *(Papio cynocephalus)*, males were less likely to fight with each other in the presence of estrus females, but they were more likely to wound each other. In other words, sexually cycling females reduced the frequency of male aggression but increased the likelihood of violent consummatory behavior. Since the frequency of social play is one of the sexually dimorphic behaviors in many primates, and since social play itself involves inhibited biting, then perhaps the effects of sexual hormones on aggression should not be looked for in differences in the frequency of aggression per se but in the modulation of high-intensity consummatory behavior by the two sexes and the situations that provoke it. If males were relatively more violent but not more frequently aggressive, their aggression would have more attention-getting power when it did occur, seriously confounding observations not done with random sampling procedures.

AGGRESSION AS CENTRIPETAL FORCE

The concept of finely tuned agonism is also compatible with the evidence from primate ranking systems. In finely tuned agonism, the part behaviors of aggression and fear replace the uninhibited execution of the consummatory acts. As primatologists have observed, the ranking systems of nonhuman primates are not characterized by episodic violence but by the directionality of threat and avoidance more akin to fine tuning. Violence occurs in these species, but it is most characteristic of either socially deprived animals or of socialized primates at the margins of familiarity. Strange male langurs will kill infants after evicting the resident male of a troop, but they are tolerant of the infants that they have sired.[7] Chimpanzees may kill strange chimpanzees from other troops, and rare deaths have also been reported from violent intergroup encounters in rhesus monkeys. Introducing a stranger to an established group of macaque monkeys reliably produces scapegoating that can result in serious injury of the newcomer (Bernstein and Mason 1963; Bernstein et al. 1974; Estrada and Estrada 1978). Similarly, in species where peaceful social relations require the spatial segregation of potential combatants, the close quarters of captivity can induce violent aggression. The hamadryas baboons studied in a zoo by Zuckerman provide a classic example (1932). In this case the proportion of adult males was so skewed that no males could establish a claim to the available females. The social system of male-female pair bonding and mutual male avoidance could not function, and the attempts of males to herd females eventually resulted in the death of the latter. Intense aggression can also be induced by placing strange monkeys together for the first time. In such situations, aggression and fear during the first hour can comprise as much as 82 percent of all social behavior, compared to an aggression rate of less than 5 percent in normal times, and such levels of agonism are about twenty times higher than that observed on subsequent days (Bernstein and Mason 1963; Bernstein 1971). Even in this case, the aggression will peak in the first few minutes of contact and drop quickly after the first hour. In macaque species, the initial aggression quickly leads to a ranking system that can be subsequently assessed by the directionality of avoidance and aggression. However, the common method of establishing social groups of macaques in captivity by the merging of separately housed individuals probably had an unfortunate effect on the development of primatological theory. As already noted, naturalistic groups of macaques are characterized by prolonged attachment relationships, familial ties, and social grooming.[8] Moreover, in baboons at least, when individual monkeys change troops, there is usually a period of habituation at the margins of the new troop. In artificially formed groups of monkeys who do not know each other, there are no mechanisms for the establishment of social relationships except those that prevail at a potentially fear-inducing social distance among total strangers. It speaks well of the inherent sociality of macaques that they form functioning social groups within a few days, but the ranking system must necessarily carry a disproportionate share of the social interaction. This is exacerbated in captivity by the concentration of food resources at particular locations that can be easily monopolized by high-ranking individuals. The aggressive consequences of concentrating resources has been documented for rhesus monkeys, and provisioning with bananas has apparently had a similar effect among the chimpanzees and baboons at Gombe Stream Reserve (V. Reynolds 1975; Southwick 1967). Group formation and resource concentration studies give clear evidence for rank effects on the distribution of resources and a positive correlation of high rank and aggressiveness. The primatological theory that developed from these observations accordingly emphasized the role of *dominance*, as it was called, as a central organizing concept in primate sociology. This concept maintained that rank differences in nonhuman primates were established by aggressiveness and that high-ranking individuals attained first access to resources. In some early studies, the method used to measure social rank was the pairing of animals in food competition tests, which necessarily created the phenomenon one sought to assess (Maslow 1936, 1940; Deag 1977).

The dominance concept made expropriation into the normal method of resource distribution in primates, and it made aggression the normal determinant of rank. Field studies of primate societies, from the 1960s on, have brought the concept of dominance into increasing disrepute, and several authors have suggested that it be abandoned entirely. However, this suggestion is too extreme, for the concept of dominance, if used judiciously, does accurately characterize some aspects of primate sociology, and the captivity studies of dominance have provided some valuable lessons. They indicate that some primates can live and breed in societies that are manifestly unnatural in the sense that the animals do not live that way if given the opportunity to do otherwise. Also, the captivity studies have shown that it is possible to create functioning social groups *de novo,* using only the extreme levels of agonism that normally prevail at the margins of social familiarity. Moreover, captivity studies have demonstrated that these groups, once established, undergo a natural evolutionary process of psychological bonding and ritualization, which makes the concept of dominance increasingly inapplicable after only one generation. Once love and ritualization have done their work, even artificially established social groups come to approximate naturalistic troops. Nonetheless, ranking systems are not simply an artifact of captivity, for linear hierarchies, based on the asymmetry of aggressive interactions, have been described for the natural troops of many primate species – *Saimiri, Pan, Macaca, Cercopithecus aethiops, Papio, Miopethecus, Propithecus,* and perhaps others—even though high-ranking individuals are not necessarily the overbearing expropriators of classical dominance theory. This being the case, the primate ranking systems in naturalistic troops cannot simply be an epiphenomenon of the prototypical aggressive encounter, analogous to artificially formed groups of animals who are strangers to each other, but must reflect a principle of primate social organization that coexists with the more affiliative forms of social behavior. The widespread geographic and taxonomic distribution of ranking systems, which occurs in New World monkeys, Old World monkeys, prosimians, and African apes, also indicates that ranking systems can arise independently of those special peculiarities of the hominid line such as a settled way of life, cooperative economic production, object exchange, and the institution of property, to which social theory has generally attributed the emergence of hierarchical societies. Since ranking relationships are manifested by aggressive behavior, regardless of their other correlates, an ethological approach would link social hierarchy primarily to the dynamics of fear and aggression and only indirectly to the distribution of rewards. If the ethological perspective is correct, then the social correlates of high rank in primates should be more closely tied to differences in aggressive role than to economic advantages, particularly in those species that have no rewards, either behavioral or material, to distribute. Moreover, since ranking systems are not derived from a war of each against all but reflect a stable principle of social organization that has been in constant interaction with other organizing principles for millions of years, social rank ought not to be regarded as a product of unsocialized aggressiveness but an epiphenomenon of the mechanism by which aggression is made social.

The ranking systems of primates are most likely an adaptation that overcomes the centripetal force of aggressive encounters.[9] Aggressive interactions usually terminate asymmetrically in the withdrawal of one of the participants, and in many situations, social dispersion is the result. As Chance and Jolly have argued (1970, p. 169),

> The trend shown by the comparative morphology of agonistic social behavior is now apparent. As the need for more persistent sociability increases it is achieved not by increasing the strength of the approach drives, but in the first stage by blocking escape by means of submission, built into an agonistic social repertoire, typified in the rat's behaviour. In the macaques and baboons this is supplemented by turning back their escape tendency towards the aggressor, so converting withdrawal into approach. This latter change, however, could not have been achieved without the provision of a submission posture for agonistic situations; the baboon and macaques possess a number, two of which are crawling up and lying in front of another animal, and sexual presentation.

Chance and Jolly are perhaps too influenced by the maternal-type bonding of adult male and female hamadryas baboons, for in maternal bonds, aggression by the care giver toward the care receiver can bond the latter even more closely. However, in most cases of agonism in macaques and baboons, fearful withdrawal is not converted into approach toward the aggressor but into aggressive approach toward a subordinate. The placatory approaches given to individuals of high rank, particularly grooming, must be kept distinct from the immediate consequences of aggressive encounters, which are not approach to the aggressor but either inhibited escape, expressed by submission, or escape followed by aggression directed to someone else. The elicitation of aggression by remote-controlled electrostimulation of the brain in primates is instructive in this regard (Delgado 1967a, 1967b; Alexander and Perachio 1973). When macaques are stimulated in aggression-producing parts of the brain, they direct their attacks against the animal immediately below them in rank, as would be expected from the widespread occurrence of redirected aggression, even though the victim is less likely to have provided "cause" than the animal just higher in rank. Aggression does function as a centripetal force in macaques and baboons, but it functions by the conversion of withdrawal into redirected aggression rather than into dispersal. The phenomenon may itself be explicable in terms of finely tuned agonism.

There is great morphological similarity between redirected aggression and the chaining of play chases that occurs in young rhesus monkeys. In play chase chaining, which the author and fellow observers dubbed *billiard ball* playing, the terminator of one chase goes on to begin another, and the encounters bounce around the social group from one monkey to the next. A similar phenomenon can also be observed in redirected aggression, where it is often called *cascaded* aggression, but here it shows more rigid directionality. Some ethologists have argued that redirection is a distinct category of behavior, which can be recognized by the fact that in ordinary aggression the behavior is directed toward the elicitor, whereas in redirection the behavior is directed toward a third party (Bastock et al. 1953). However, there are many cases in which aggression is directed against so-called third parties in which the postulation of substitute objects is counterintuitive. In pain-induced aggression, for example, the animal may strike out against a social target even though the aggression is provoked by an electric shock from the floor. Squirrel monkeys, for example, if given an electric shock, will solve a problem to open a door to get access to an animate target (Azrin et al. 1965). If the fine tuning of agonistic behavior implies the maintenance of a delicate balance between aggression and fear, so neither system can control with full intensity, then withdrawal from a fear-inducing situation may well tip the balance in favor of aggression and allow expression toward a lower-ranking target, with any object substitutes being in the mind of the observer and not in the mind of the animal.

The present analysis suggests that ranking systems are the product of finely tuned agonism among socialized animals in a social context. Consequently, it is important to distinguish *ranking systems,* which channel aggression centripetally, from mere agonistic interactions in confined spaces where dispersal effects are impossible to observe. Since all primates are capable of aggressive behavior, the difference between mere asymmetric termination and rank-related behavior has to do with the social consequences of the action. The capacity for true ranking systems apparently does not occur in all primates, even in those who can form dominant-subordinate relationships in captivity, like the *Cercopithecus* monkeys observed by Rowell (1971). Rather, it is most likely an adaptation for large social groups of otherwise aggressive animals, whose group boundaries enclose individuals not related through the developmental attachment channels of maternal and filial love, with their concomitant inhibited aggression. The rhesus and Japanese macaques in fact provide good examples of this principle, because their societies consist of multiple solidary matrilines, each composed of the adult female descendants of single females and any of their dependent offspring; the matrilines themselves are ranked in linear hierarchies within the troop; and entire troops may be linearly ranked at points of home range overlap. Human beings are the only other mammals that have such complex social organization with an equivalent ability to integrate strangers into the system. However, hominids have carried Chance and Jolly's agonistic progression one step further: they have not only retained an ability for submission and redirected aggression but have elevated inhibited escape into a rewarding consummatory activity as indicated by the analysis of the smile.

An ethological analysis of primate aggression marks the end of the war of each against all, for it makes aggression the major mechanism of social affiliation as well as of social dispersal. Second, an ethological analysis, by regarding ranking systems as a phylogenetically progressive adaptation to increased sociality, repudiates the idea that aggression is a property of the atomistic individual before the social contract. Primate ranking systems have probably been in existence for many millions of years, and it is likely that they antedate the emergence of the genus *Homo*. Consequently, ranking systems should be closely interwoven with other organizational principles of primate society and the acquisition of status should be only partially dependent upon aggressive superiority. Studies of macaques spanning five monkey generations have shown conclusively that social rank is not necessarily achieved at all (Gouzoules 1975; Koford 1963; Koyama 1967; Missakian 1972). In animals that remain in their natal troops—in macaque species almost always females—social rank is not achieved but ascribed. It is inherited, in a social sense of the term, from one's mother. Moreover, since social rank is also highly age dependent, most individuals rise in rank simply by getting older. It is primarily the young adult males who must achieve their status by migrating to a strange troop and fighting with resident males. Even here, however, the new males are integrated into the existing male hierarchy and the resident males are not evicted, as occurs in many species of primates with unimale troops. Nor can aggression and rank be considered a property of the atomistic individual in isolation from its fellows. Recent studies show that the status competition of migrating males can be a cooperative venture, in which males are aided by their brothers who migrate with them or join the same troop at a later date. Coalitions among the resident males against newcomers were also recorded in one of the first field studies of savanna baboons (Riss and Goodall 1977; De Waal 1978; Hall and DeVore 1965). In an artificial formation of a rhesus monkey troop, the females also forged alliances with particular males and instigated them to attack other females (Bernstein, Gordon, and Rose 1974). In hamadryas baboons, the phenomenon of tripartite threat has also been recorded, in which a low-ranking animal positions itself between two higher ranking animals and threatens the lower of the two. The target of the attack cannot easily reciprocate because it would necessarily be threatening in the direction of an animal that outranked it. The occurrence of coalition behaviors provides the background to Bernstein's suggestion that rank-oriented primates have not been selected for increased aggressiveness at all but for the political and social skills needed to forge coalitions instrumental in status competition. Although seemingly a simple proposition, it is deceptively simple, for if taken seriously it implies a far different psychology from that which has hitherto prevailed. It implies the capacity to utilize the mechanisms of affiliation in the service of aggression in exactly the same way as aggression is used in the service of attachment.

Coalition and competition are two aspects of the same underlying psychological capacity: the instrumental control of affect and the affective control of instrumentality.

SUMMARY

Ethological study of primate societies indicates that aggression and fear are systematically related to the boundaries of social attachment. Agonism is used both to terminate attachment relationships and to repel potential threats to an existing relationship. Moreover, attachment figures serve to inhibit the reactivity of both aggression and fear during development. This inhibited reactivity in turn provides the mechanism for social relationships derived from inhibited biting (play), inhibited escape, and redirected aggression (ranking systems). The systematic relationships among agonistic behaviors, social group boundaries, and social experience indicate that these systems are among the most important *social* mechanisms primates possess. Moreover, the systematic relationship between attachment, agonism, and the assessment of novelty indicates that these mechanisms are integrated with cognitive models of the world and cannot be reduced to the innate behavior patterns that implement them. Rather, the evidence is supportive of phylogenetically progressive limbic forebrain mechanisms that assess perceived threats to conceptualized social relationships and program the appropriate agonistic response through the modulation of threshold levels of fixed-action patterns and the motivation of instrumental action. Attachments themselves are stabilized approach-avoidance conflicts.

NOTES TO CHAPTER 3

1. See references in this section and Stern et al. (1975) and Condon and Sander (1974).

2. Ahrens cited by Vine (1973); also Eibl-Eibesfeldt (1975, p. 449), Gewirtz (1965).

3. Refer to chap. 2.

4. See also Anderson and Mason (1974).

5. See chap. 4.

6. See references to chap. 2.

7. See references in P. Reynolds (1976).

8. Discussed in P. Reynolds (1976) and also in chap. 5.

9. This should not be confused with the hypothesis that rank *reduces* aggression in a social group (Tinbergen 1953, p. 71). For an overview of primate aggression see Holloway (1974).

REFERENCES

ALEXANDER, M., and PERACHIO, A. A. 1973. The influence of target sex and dominance on evoked attack in rhesus monkeys. *Amer. J. Phys. Anthro.* 38:543–547.

ALTMANN, STUART A. 1961. The social play of rhesus monkeys. *American Zoologist* 1, no. 4:27, abstract.

1962. A field study of the sociobiology of rhesus monkeys, *Macaca mulatta. Annals of N.Y. Acad. of Sciences* 102, pt. 2: 338–435.

ANDERSON, C. O., and MASON, W. A. 1974. Early experience and complexity of social organization in groups of young rhesus monkeys *(Macaca mulatta). J. Comp. Physiol. Psychol.* 87:681–690.

ANDREW, R. J. 1964. The displays of the primates. In *Evolutionary and genetic biology of primates*, vol. 2, ed. J. Buettner-Janusch, pp. 227–309. New York: Academic Press.

ARGYLE, MICHAEL, and COOK, MARK 1976. *Gaze and mutual gaze.* Cambridge: Cambridge U. Press.

ARLING, G. L., and HARLOW, H. F. 1967. Effects of social deprivation on maternal behavior of rhesus monkeys. *J. Comp. Physiol. Psychol.* 64:371–377.

AZRIN, N. H.; HUTCHINSON, R. R.; and McLAUGHLIN, R. 1965. The opportunity for aggression as an operant reinforcer during aversive stimulation. *J. Exp. Analysis of Behavior* 8:171–180.

BASTOCK, M.; MORRIS, D.; and MOYNIHAN, M. 1953. Some comments on conflict and thwarting in animals. *Behavior* 6:66–84.

BATESON, GREGORY 1955. A theory of play and fantasy. *A.P.A. Psychiatric Reports 2.* Reprinted in *Steps toward an ecology of mind,* pp. 177–193. New York: Ballantine, 1972.

BEKOFF, MARC 1972. The development of social interaction, play and metacommunication in mammals: an ethological perspective. *Quarterly Rev. Biol.* 47:412–434.

1977. Social communication in canids: evidence for the evolution of a stereotyped mammalian display. *Science* 197:1097–1099.

BERNSTEIN, I. S. 1970. Activity patterns in pigtail monkey groups. *Folia Primatol.* 12:187–198.

1971. Activity profiles of primate groups. In *Behavior of nonhuman primates*, vol. 3, ed. Allan Schrier and Fred Stollnitz. New York: Academic Press.

BERNSTEIN, IRWIN S., and GORDON, THOMAS P. 1974. The function of aggression in primate societies. *American Scientist* 62:304–311.

BERNSTEIN, I. S.; GORDON, T. P.; and ROSE, R. M. 1974. Aggression and social controls in rhesus monkey *(Macaca mulatta)* groups revealed in group formation studies. *Folia Primatol.* 21:81–107.

BERNSTEIN, I. S., and MASON, W. A. 1963. Group formation by rhesus monkeys. *Animal Behaviour* 11:28–31.

1970. Effects of age and stimulus conditions on the emotional responses of rhesus monkeys: differential responses to frustration and to fear stimuli. *Developmental Psychobiology* 3:5–12.

BLEST, A. D. 1957. The function of eyespot patterns in Lepidoptera. *Behaviour* 11:209–255.

BLURTON-JONES, N. G. 1968. Observations and experiments on causation of threat displays of the great tit *(Parus major)*. *Animal Behaviour. Monograph 1*, no. 2.

BOWLBY, JOHN 1958. The nature of the child's tie to his mother. *Internat'l J. Psychoanalysis* 39:350–373.

BRANDT, E. M.; STEVENS, C. W.; and MITCHELL, G. 1971 Visual and social communication in adult male isolate-reared monkeys *(Macaca mulatta)*. *Primates* 12:105–112.

BRANNIGAN, CHRISTOPHER R., and HUMPHRIES, DAVID A. 1972. Human non-verbal behaviour, a means of communication. In *Ethological studies of child behaviour*, ed. N. Blurton-Jones, pp. 37–64. Cambridge: Cambridge U. Press.

BRAZELTON, T. BERRY; TRONICK, EDWARD; ADAMSON, LAUREN; ALS, HEIDELISE; and WISE, SUSAN 1975. Early mother-infant reciprocity. *Ciba Foundation Symposium*, no. 33, n.s.:137–154. Amsterdam: Elsevier.

BRONSON, GORDON W. 1968. The development of fear in man and other animals. *Child Development* 39:409–431.

CHANCE, M. R. A., and JOLLY, C. J. 1970. *Social groups of monkeys, apes and men.* London: Jonathan Cape.

CHARLESWORTH, WILLIAM R., and KREUTZER, MARY ANNE 1973. Facial expressions of infants and children. In *Darwin and facial expression: a century of research in review*, ed. Paul Ekman, pp. 91–168. New York: Academic Press.

CHEVALIER-SKOLNIKOFF, SUZANNE 1973. Facial expression of emotion in nonhuman primates. In *Darwin and facial expression*, ed. Paul Ekman, pp. 11–89. New York and London: Academic Press.

1974. The primate play face: a possible key to the determinants and evolution of play. *Rice U. Studies* 60, no. 3:9–29.

CONDON, WILLIAM S., and SANDER, LOUIS W. 1974. Neonate movement is synchronized with adult speech: interactional participation and language acquisition. *Science* 183:99–101.

COSS, R. G. 1970. The perceptual aspects of eye-spot patterns and their relevance to gaze behaviour. In *Behaviour Studies in Psychiatry*, ed. S. J. Hutt and Corinne Hutt, pp. 121–147. Oxford: Pergamon Press.

DARWIN, CHARLES 1965. *The expression of the emotions in man and animals.* Chicago and London: U. of Chicago Press. 1st ed., 1872.

DAWKINS, RICHARD 1976. Hierarchical organisation: a candidate principle for ethology. In *Growing points in ethology*, ed. P. P. G. Bateson and R. A. Hinde, pp. 7–54. Cambridge: Cambridge U. Press.

DEAG, JOHN M. 1977. Aggression and submission in monkey societies. *Animal Behaviour* 25:465–474.

DELGADO, JOSE M. R. 1967a. Aggression and defense under cerebral radio control. In *Aggression and defense: neural mechanisms and social patterns*, ed. C. D. Clemente and D. B. Lindsley, pp. 171–193. Berkeley and Los Angeles: U. of California Press.

1967b. Social rank and radio-stimulated aggressiveness in monkeys. *J. Nerv. Ment. Dis.* 144:383–390.

DE WAAL, FRANS B. M. 1978. Exploitative and familiarity dependent support strategies in a colony of semi-free living chimpanzees. *Behaviour* 67: 268–312.

EIBL-EIBESFELDT, IRENAUS 1971. *Love and hate.* Trans. Geoffrey Strachen. London: Methuen. German ed., 1970.

1972. Similarities and differences between cultures in expressive movements. In *Nonverbal communication*, ed. R. A. Hinde, pp. 297–314. Cambridge: Cambridge U. Press.

1973. The expressive behaviour of the deaf-and-blind born. In *Social communication and movement*, ed. M. van. Cranach and I. Vine, pp. 163–194. New York: Academic Press.

1975. *Ethology: the biology of behavior.* 2nd ed. New York: Holt, Rinehart and Winston.

EKMAN, PAUL 1973. Cross-cultural studies of facial expression. In *Darwin and facial expression: a century of research in review*, ed. P. Ekman, pp. 169–222. New York and London: Academic Press.

1977. Biological and cultural contributions to body and facial movement. In *The anthropology of the body*, ASA Monograph No. 15, ed. John Blacking, pp. 39–84. London and New York: Academic Press.

EKMAN, PAUL, and FRIESEN, W. V. 1971. Constants across cultures in the face and emotion. *Personality and Soc. Psych.* 17, no. 2:124–129.

ESTRADA, A., and ESTRADA, R. 1978. Changes in social structure and interactions after the introduction of a second group in free-ranging troop of stumptail macaques *(Macaca arctoides):* social relations II. *Primates* 19: 665–680.

FINCH, G. 1945. Chimpanzee frustration responses. *Psychosomatic Medicine* 4: 233–251.

FISCHER, H. 1965. Das triumphgeschrei der graugans *(Anser anser)*. *Z. Tierpsychol.* 22:247–304.

FLYNN, J. P. 1967. The neural basis of aggression in cats. *Neurophysiology and emotion,* ed. D. C. Glass, pp. 40–60. New York: Rockefeller U. Press and Russell Sage Foundation.

FOLEY, J. P., Jr. 1935. Judgement of facial expression of emotion in the chimpanzee. *J. Soc. Psychol.* 6:31–67.

FREEDMAN, D. 1965. Hereditary control of early social behavior. In *Determinants of infant behavior,* ed. B. M. Foss, vol. 3, pp. 149–159. London: Methuen.

GALLUP, GORDON G., Jr. 1975. Towards an operational definition of self-awareness. In *Socioecology and psychology of primates,* ed. R. H. Tuttle, pp. 309–341. The Hague: Mouton.

GARDNER, B. T., and WALLACH, L. 1965. Shapes of figures identified as a baby's head. *Perceptual and Motor Skills* 20:135–142.

GARFINKEL, HAROLD 1967. *Studies in ethnomethodology.* Englewood Cliffs, N.J.: Prentice-Hall.

GASTAUT, H., and BERT, J. 1969. Electroencephalographic detection of sleep induced by repetitive sensory stimuli. In *On the nature of sleep,* ed. G. E. Wolstenholme et al, pp. 260–283. London: Churchill.

GAZZANIGA, MICHAEL S. 1970. *The bisected brain.* New York: Appleton-Century-Crofts.

GESELL, A., ed. 1940. *The first five years of life: a guide to the study of the preschool child.* London: Methuen.

GEWIRTZ, J. L. 1965. The course of infant smiling in four child-rearing environments in Israel. In *Determinants of infant behaviour,* ed. B. M. Foss, vol. 3, pp. 205–248. London: Methuen.

GOLDFOOT, DAVID A. 1977. Sociosexual behaviors of nonhuman primates during development and maturity: social and hormonal relationships. In *Behavioral primatology: advances in research and theory,* ed. Allen M. Schrier, vol. 1, pp. 139–184. Hillsdale, N.J.: Lawrence Erlbaum.

GOLDMAN, PATRICIA S., et al. 1974. Sex-dependent behavioral effects of cerebral cortical lesions in the developing rhesus monkey. *Science* 186:540–542.

GOUZOULES, H. 1975. Maternal rank and early social interactions of infant stumptail macaques, *Macaca arctoides. Primates* 16:405–418.

HALL, K.R.L. 1962. Behaviour of monkeys toward mirror-images. *Nature* (London) 196:1258–2161.

HALL, K. R. L., and DeVORE, IRVEN 1965. Baboon social behavior. In *Primate behavior: field studies of monkeys and apes,* ed. Irven DeVore, pp. 53–110. New York: Holt, Rinehart and Winston.

HANSEN, ERNST W. 1966. The development of maternal and infant behavior in the rhesus monkey. *Behaviour* 27:107–149.

HARLOW, H. F.; DODSWORTH, R. O.; and HARLOW, M. K. 1965. Total social isolation in monkeys. *Proc. Nat. Acad. Sciences (Washington)* 54:90–97.

HARLOW, HARRY, and HARLOW, MARGARET 1965. The affectional system. In *Behavior of nonhuman primates,* ed. A. Schrier, H. Harlow, and F. Stollnitz, vol. 2, pp. 287–334. New York: Academic Press.

HARLOW, H. F.; HARLOW, M. K.; SCHILTZ, K. A.; and MOHR, D. J. 1971. The effect of early adverse and enriched environments on the learning ability of rhesus monkeys. In *Cognitive processes of nonhuman primates,* ed. L. E. Jarrard, pp. 121–148. New York: Academic Press.

HARLOW, H. F., and SUOMI, S. J. 1970. Nature of love—simplified. *American Psychologist* 25:161–168.

HAUSFATER, GLENN 1975. Dominance and reproduction in baboons *(Papio cynocephalus):* a quantitative analysis. *Contributions to Primatology* 7. Basel: Karger.

HEBB, D. O. 1945. The forms and conditions of chimpanzee anger. *Bulletin of Canadian Psychological Assoc.* 2:32–35.

1961. *The organization of behavior.* New York: Wiley, 1st ed., 1949.

HINDE, R. A. 1961. The establishment of the parent-offspring relation in birds, with some mammalian analogies. In *Current problems in animal behavior,* ed. W. H. Thorpe and O. L. Zangwill, pp. 175–193. Cambridge: Cambridge U. Press.

HINDMARCH, IAN 1973. Eyes, eye-spots and pupil dilation in non-verbal communication. In *Social communication and movement,* ed. M. van Cranach and Ian Vine, pp. 299–321. New York: Academic Press.

HOFER, MYRON A. 1978. Hidden regulatory processes in early social relationships. In *Perspectives in ethology,* ed. P. P. G. Bateson and P. H. Klopfer, vol. 3, pp. 135–166. New York and London: Plenum Press.

HOLLOWAY, RALPH L., ed. 1974. *Primate aggression, territoriality, and xenophobia: a comparative perspective.* New York and London: Academic Press.

HRDY, SARAH BLAFFER 1977. Infanticide as a primate reproductive strategy. *American Scientist* 65:40–49.

IZARD, C. E. 1971. *The face of emotion.* New York: Appleton.

KAADA, BIRGER 1967. Brain mechanisms related to aggressive behavior. In *Aggression and defense: neural mechanisms and social patterns,* ed. C. D. Clemente and D. B. Lindsley, pp. 95–133. Berkeley and Los Angeles: U. of California Press.

KAGAN, JEROME 1978. The enhancement of memory in infancy. *Quarterly Newsletter of the Institute for Comparative Human Development* 2:58–60.

KAUFMAN, I. C., and ROSENBLUM, L. A. 1967. The reaction to separation in infant monkeys: anaclitic depression and conservation-withdrawal. *Psychosomatic Medicine* 29:648–676.

KENDON, ADAM 1975. Some functions of the face in a kissing round. *Semiotica* 15:299–334.

KLING, ARTHUR, and MASS, ROSLYN 1974. Alterations of social behavior with neural lesions in nonhuman primates. In *Primate aggression, territoriality, and xenophobia: a comparative perspective,* ed. R. L. Holloway, pp. 361–386. New York and London: Academic Press.

KLING, A., and STEKLIS, H. D. 1976. Neural substrate for affiliative behavior in nonhuman primates. *Brain, Behaviour and Evolution* 13:216–238.

KOFORD, C. B. 1963. Rank of mothers and sons in bands of rhesus monkeys. *Science* 141:356–357.

KOYAMA, N. 1967. On dominance rank and kinship of a wild Japanese monkey troop in Arashiyama. *Primates* 8:189–216.

KRUIJT, J. P. 1964. Ontogeny of social behaviour in Burmese red jungle-fowl (*Gallus gallus spadiceus* Bonnaterre). *Behaviour,* supplement 12.

KUMMER, HANS 1968. *The social organization of hamadryas baboons: a field study.* Basel: Karger.

LA BARRE, WESTON 1947. The cultural basis of emotions and gestures. *J. Personality* 16:49–68.

LAWICK-GOODALL, JANE van 1968. A preliminary report on expressive movements and communication in the Gombe Stream chimpanzees. In *Primates: adaptation and variability,* ed. P. Jay, pp. 313–374. New York: Holt, Rinehart and Winston.

LEACH, EDMUND 1972. The influence of cultural context on non-verbal communication in man. In *Nonverbal communication,* ed. R. A. Hinde, pp. 315–347. Cambridge: Cambridge U. Press.

LEVY, J.; TREVARTHEN, C.; and SPERRY, R. W. 1972. Perception of bilateral chimeric figures following hemisphere deconnection. *Brain* 95:61–78.

LEYHAUSEN, PAUL 1973. On the function of the relative hierarchy of moods (as exemplified by the phylogenetic and ontogenetic development of prey-catching in carnivores). In *Motivation of human and animal behavior: an ethological view,* ed. K. Lorenz and P. Leyhausen, pp. 144–247. New York: Van Nostrand Reinhold. German ed., 1965.

MacLEAN, PAUL D. 1964. Mirror display in the squirrel monkey, *Saimiri sciureus. Science* 146:950–952.

MARLER, P. 1976. On animal aggression: the roles of strangeness and familiarity. *American Psychologist* 30:239–246.

MASLOW, A. H. 1936. The role of dominance in the social behavior of infrahuman primates. IV. The determination of hierarchy in pairs and in a group. *J. Genetic Psychol.* 49:161–198.

1940. Dominance quality and social behavior in infrahuman primates. *J. Soc. Psychol.* 11:313–324.

MASON, WILLIAM A. 1963. The effects of environmental restriction on the social development of rhesus monkeys. In *Primate social behavior,* ed. Charles H. Southwick, pp. 161–173. Princeton, N.J.: Van Nostrand.

1973. Regulatory functions of arousal in primate psychosexual development. In *Behavioral regulators of behavior in primates,* ed. C. R. Carpenter, pp. 19–33. Lewisberg: Bucknell U. Press.

MASON, W. A., and BERKSON, G. 1962. Conditions influencing vocal responsiveness of infant chimpanzees. *Science* 137:127–128.

MAURER, D., and SALAPATEK, P. 1976. Developmental changes in the scanning of faces by infants. *Child Development* 47:523–527.

MELZACK, R. 1965. Effects of early experience on behavior: experimental and conceptual considerations. In *Psychopathology of perception,* ed. P. H. Hock and J. Zubin. New York: Grune and Stratton.

MELTZOFF, A. N., and MOORE, M. K. 1977. Imitation of facial and manual gestures by human neonates. *Science* 198:75–78.

MILLER, R. E. 1971. Experimental studies of communication in the monkey. *Primate Behaviour* 2:139–175.

MILLER, ROBERT E.; BANKS, JAMES H., Jr.; and OGAWA, NOBUYA 1963. Role of facial expression in "cooperative avoidance conditioning" in monkeys. *J. Abnormal and Soc. Psych.* 67:24–30.

MISSAKIAN, E. A. 1972. Genealogical and cross-genealogical dominance relations in a group of free-ranging rhesus monkeys *(Macaca mulatta)* on Cayo Santiago. *Primates* 13:169–180.

MITCHELL, G. D. 1968. Persistent behavior pathology in rhesus monkeys following early social isolation. *Folia Primatol.* 8:132–147.

1970. Abnormal behavior in primates. In *Primate behavior: developments in field and laboratory research,* ed. L. A. Rosenblum, vol. 1, pp. 195–249. New York and London: Academic Press.

MONEY, JOHN, and EHRHARDT, ANKE A. 1972. *Man and woman, boy and girl.* Baltimore and London: Johns Hopkins Press.

MONRAD-KROHN, G. H. 1924. On the dissociation of voluntary and emotional innervation in facial paresis of central origin. *Brain* 47:22–35.

MYERS, R. E. 1969. Neurology of social communication in primates. *Proc. Second Int. Congr. Primat.* 3:1–9. Basel and New York: Karger.

1972. Role of prefrontal and anterior temporal cortex in social behavior and affect in monkeys. *Acta Neurobiol. Exp.* 32:567–579.

PANSKEPP, JAAK 1971a. Effects of hypothalamic lesions on mouse-killing and shock-induced fighting in rats. *Physiology and Behavior* 6:311–316.

1971*b*. Drugs and stimulus-bound attack. *Physiology and Behavior* 6:317–320.

1971*c*. Aggression elicited by electrical stimulation of the hypothalamus of albino rats. *Physiology and Behaviour* 6:321–329.

PAPOUSEK, H., and PAPOUSEK, M. 1975. Cognitive aspects of preverbal social interactions between human infants and adults. *Ciba Foundation Symposium*, no. 33, N.S.: 241–269. Amsterdam: Elsevier.

PETRINOVICH, LEWIS 1973. Darwin and the representative expression of reality. In *Darwin and facial expression: a century of research in review*, ed. Paul Ekman, pp. 223–256. New York: Academic Press.

POECK, K. 1969. Pathophysiology of emotional disorders associated with brain damage. In *Handbook of clinical neurology. Vol. 3. Disorders of higher nervous activity*, ed. P. J. Vinken and G. W. Bruyn, pp. 343–367. Amsterdam: North-Holland; New York: Wiley Interscience.

POMPEIANO, O. and SWETT, J. E. 1962*a*. EEG and behavioral manifestations of sleep induced by cutaneous nerve stimulation in normal cats. *Archives Italiennes de Biologie* 100:311–342.

1962*b*. Identification of cutaneous and muscular afferent fibers producing EEG synchronization or arousal in normal cats. *Archives Italienne de Biologie* 100:343–380.

POOLE, TREVOR B. 1978. An analysis of social play in polecats *(Mustelidae)* with comments on the form and evolutionary history of the open mouth play face. *Animal Behaviour* 26:36–49.

PRECHTL, H. F. R. 1958. The directed head turning response and allied movements of the human baby. *Behaviour* 13:212–242.

PRIBRAM, KARL H. 1961. Limbic system. In *Electrical stimulation of the brain*, ed. Daniel E. Sheer, pp. 311–320. Austin: U. of Texas Press.

1967. Emotion: steps toward a neuropsychological theory. In *Neurophysiology and emotion*, ed. D. C. Glass, pp. 3–40. New York: Rockefeller U. Press and Russell Sage Foundation.

1969. The neurobehavioral analysis of limbic forebrain mechanisms. In *Advances in the study of behavior.* ed. Daniel S. Lehrman, Robert A. Hinde, and Evelyn Shaw, vol. 2, pp. 297–332. New York: Academic Press.

Forthcoming. Emotions. In *Handbook of Clinical Neuropsychology*, ed. S. B. Filskov and J. J. Boll.

RANDOLPH, M. C., and MASON, W. A. 1969. Effects of rearing conditions on distress vocalizations in chimpanzees. *Folia Primatol.* 10:103–112.

REDICAN, WILLIAM K. 1975. Facial expressions in nonhuman primates. *Primate behavior: developments in field and laboratory research*, ed. Leonard Rosenblum, vol. 4, pp. 104–194. New York: Academic Press.

REYNOLDS, PETER C. 1976. The emergence of early hominid social organization. I. The attachment systems. *Yrbk. Physical Anthropology* 20:73–95.

REYNOLDS, VERNON 1975. How wild are the Gombe chimpanzees? *Man* n.s. 10:123–125.

RISS, DAVID, and GOODALL, JANE 1977. The recent rise to the alpha-rank in a population of free-living chimpanzees. *Folia Primatol.* 27:134–151.

ROBERTS, W. W., and KIESS, H. O. 1964. Motivational properties of hypothalamic agression in cats. *J. Comp. and Physiol. Psych.* 58:187–193.

ROBSON, KENNETH S. 1967. The role of eye-to-eye contact in maternal-infant attachment. *J. Child Psychol. Psychiat.* 8:13–25.

ROSENBLUM, LEONARD A. 1971. The ontogeny of mother-infant relations in macaques. In *The ontogeny of vertebrate behavior*, ed. Howard Moltz, pp. 315–367. New York: Academic Press.

   1974. Sex differences in mother-infant attachment in monkeys. In *Sex differences in behavior,* ed. R. C. Friedman, R. M. Richart, and R. L. Van de Wiele, pp. 123–141. New York: Wiley.

ROSVOLD, H. E.; MIRSKY, A. F.; and PRIBRAM, K. H. 1954. Influence of amygdalectomy on social interaction in a monkey group. *J. Comp. and Physiol. Psychol.* 47:173–178.

ROWELL, T. E. 1971. Organization of caged groups of *Cercopithecus* monkeys. *Animal Behaviour* 19:625–645.

SACKETT, G. P. 1966. Monkeys reared in isolation with pictures as visual input: evidence for an innate releasing mechanism. *Science* 154:1468–1473.

   1972. Exploratory behavior of rhesus monkeys as a function of rearing experiences and sex. *Developmental Psychol.* 6:260–270.

   1974. Sex differences in rhesus monkeys following varied rearing experiences. In *Sex differences in behavior,* ed. R. C. Friedman, R. M. Richart, and R. L. Van de Wiele, pp. 99–121. New York: Wiley.

SACKETT, G. P.; HOLM, R. A.; and RUPPENTHAL, GERALD C. 1976. Social isolation rearing: species differences in behavior of macaque monkeys. *Developmental Psychol.* 12:283–288.

SALZEN, E. A. 1967. Imprinting in birds and primates. *Behaviour* 28:232–254.

SCHNEIRLA, T. C. 1959. *An evolutionary and developmental theory of biphasic processes underlying approach and withdrawal.* Nebraska Symposium on Motivation. Lincoln: University of Nebraska Press. Reprinted in *Principles of animal psychology,* ed. N. R. F. Maier and T. C. Schneirla, pp. 511–554. New York: Dover, 1964.

SCHREINER, LEON, and KLING, ARTHUR 1953. Behavioral changes following rhinencephalic injury in cat. *J. Neurophysiology* 16:643–699.

SLOTNIK, B. M. 1967. Disturbances of maternal behavior in the rat following lesions of the cingulate cortex. *Behaviour* 29:204–236.

SLOTNIK, B. M., and NIGOSH, B. J. 1975. Maternal behavior of mice with cingulate, cortical, amygdala, or septal lesions. *J. Comp. Physiol. Psychol.* 88:118–127.

SOUTHWICK, CHARLES H. 1967. An experimental study of intragroup agonistic behavior in rhesus monkeys *(Macaca mulatta)*. *Behaviour* 28:182–209.

SPERRY, R. W. 1974. Lateral specialization in the surgically separated hemispheres. In *The neurosciences: third study program*, ed. F. O. Schmitt and F. G. Worden, pp. 5–19. Cambridge, Mass.: MIT Press.

STERN, D. N.; JAFFE, J.; BEEBE, B.; and BENNETT, S. L. 1975. Vocalizing in unison and alternation: two modes of communication within the mother-infant dyad. *Annals N.Y. Acad. of Sciences* 263:89–100.

SUOMI, S. J., and HARLOW, H. F. 1972. Social rehabilitation of isolate-reared monkeys. *Developmental Psychol.* 6:487–496.

TEUBER, HANS-LUKAS 1975. Effects of focal brain injury on human behavior. In *The nervous system*, ed. D. B. Tower, vol. 3, pp. 457–480. New York: Raven Press.

TINBERGEN, N. 1953. *Social behaviour in animals*. London: Methuen.

TOMKINS, SILVAN S. 1962. *Affect, imagery, consciousness.* 2 vols. New York: Springer.

TREVARTHEN, COLWYN 1977. *Instincts for human understanding and for cultural cooperation: their development in infancy.* Paper presented at Werner-Reimer-Stiftung Symposium on Human Ethology, Bad Homburg, West Germany, Oct. 25–29.

VAN HOOFF, J. A. R. A. M. 1967. The facial displays of Catarrhine monkeys and apes. In *Primate Ethology*, ed. D. Morris, pp. 7–68. Chicago: Aldine.

1973. A structural analysis of the social behaviour of a semi-captive group of chimpanzees. In *Social communication and movement*, ed. M. van Cranach and I. Vine, pp. 75–162. New York: Academic Press.

VINE, IAN 1973. The role of facial-visual signalling in early social development. In *Social communication and movement*, ed. M. van Cranach and I. Vine, pp. 195–298. New York: Academic Press.

WASER, P. M. 1975. Experimental playbacks show vocal mediation of intergroup avoidance in a forest monkey. *Nature* (London) 255:56–58.

WASHBURN, R. W. 1929. A study of the smiling and laughing of infants in the first year of life. *Genetic Psychol. Monographs* 6:397–535.

WHITE, B. L., and HELD, R. 1966. Plasticity of sensori-motor development in the human infant. In *The causes of behavior—readings in child development and educational psychology*, 2nd ed., ed. J. F. Rosenblith and W. Allinsmith, pp. 60–70. Boston: Bacon.

WICKLER, WOLFGANG 1968. *Mimicry in plants and animals.* Trans. R. D. Martin. New York and Toronto: McGraw-Hill.

WOLFF, PETER 1963. Observations on the early development of smiling. In *Determinants of infant behaviour,* ed. B. M. Foss, vol. 2, pp. 113–138. London: Methuen; New York: Wiley.

YERKES, R. M., and YERKES, A. W. 1970. *The great apes: a study of anthropoid life.* New York: Johnson Reprint. 1st ed., 1929.

YIN, R. K. 1970. Face-recognition by brain-injured patients: a dissociable disability? *Neuropsychologia* 8:395–402.

ZUCKERMAN, SOLLY 1932. *The social life of monkeys and apes.* London: Kegan Paul.

IN THE BEGINNING
WAS THE CONCEPT

Although affective goals are innately linked to instrumental actions like biting, such innate components coexist in primates with a progressive elaboration of brain mechanisms capable of abstracting the objective consequences of action and storing them in a form that is accessible to mental manipulation. An information store abstracted through individual experience, capable of assimilating novel content through processes of induction, and serving as a basis for prediction through the process of deduction is generally called *conceptual*.[1] How such conceptual information storage actually works is one of the major unsolved problems of psychology, but more than a half century of experimentation and observation have conclusively shown that nonhuman primates nonetheless can process information in this way.

The proof that animals can form concepts requires systematic testing in which simpler strategies have been eliminated by the experimental design. For example, the ability to recognize and remember a particular stimulus is not necessarily conceptual, as it may reflect a conditional reaction, like a dog that barks at a person who once kicked it. Since conditioning is distributed throughout the vertebrate world, the postulation of distinct forms of cognitive processing requires controls to eliminate the more common alternatives. Conditioned reactions generalize to novel stimuli, but the conditioned reaction decreases as the stimulus becomes more dissimilar to the training stimulus. In conceptual processing, on the other hand, the training can generalize to entirely new classes of stimuli, which are totally dissimilar to the training stimuli on purely physical criteria.

Moreover, in relational concepts, such as "the third in a series" or "larger than," the conceptual category is independent of the perceptual properties of particular stimuli used. In many naturalistic situations, it will be impossible to tell whether conceptual processing ought to be inferred, but in the laboratory it is possible to test animals in such a way that all simpler explanations are eliminated. The strongest proof of conceptual processing is provided when an animal can discriminate a concept exemplar from a nonexemplar, even when it has never seen either stimulus object before and the stimulus features are in fact irrelevant to the concept definition. Primates, in fact, can carry the rigorous criterion one step further by choosing a concept exemplar correctly from among novel stimuli, even though the latter are presented in a different sensory modality (such as tactile instead of visual) and with a different number of physical dimensions (two instead of three).

Operationally, at least, it is important to distinguish sensory concepts from purely relational ones. In sensory concepts the relationships that are abstracted and encoded refer to aspects of the perceptual image itself. Home-reared chimpanzees (and pigeons, too) can recognize pictures of people and discriminate people from other classes of perceptual phenomena. However, it would not be unreasonable to suppose that the concept "people" had associated with it certain perceptual recognition criteria that differed from the recognition criteria for, say, "apples." Many of the geometric figures used in tests of conceptualization suggest this, as simple alterations in some of the perceptual attributes can change it from one kind of figure into another, as when a square is changed to a rhombus by altering the internal angles. Sensory concepts, such as circle and square, are often used to test cross-modal abilities in primates, as the constancy of internal relationships may be unaffected by transposition to another modality (Cowey and Weiskrantz 1975; Ettlinger and Jarvis 1976; Davenport and Rogers 1970; Davenport et al. 1973; Rogers and Davenport 1975). Davenport and Rogers, for example, used the fact that the internal relationships of an object are the same in both visual and tactile modalities to demonstrate cross-modal matching-to-sample in anthropoid apes. The animals were first shown an object (the sample). Next the sample and a second object were placed behind a screen, and the ape had to retrieve, by tactile manipulation alone, the object it had just seen. Davenport and Rogers went on to perform the converse experiment and show that chimpanzees could choose a tactilely mediated *(haptic)* sample object from two test objects presented only visually. These experiments are theoretically important, because previous experiments with nonhuman primates using other methods had failed to show any such effect, and the supposed absence of cross-modal matching had been postulated as the reason why nonhuman primates lacked language. Chimpanzees can also retain the mental encoding of the sample stimulus for at least up to twenty seconds, as tested on a delayed response variant of the same procedure, and their performance is unaffected by the substitution of a color photograph of an object for the sample object itself. Black and white photographs, however, produced a slight decrement in performance. These experiments demonstrate that great apes can abstract relational uniformities from sensory information in one modality and use this same relational information to interpret perceptual information in another modality. Moreover, the relational information can still be abstracted and used in spite of transpositions in size, color, and dimensionality. However, high-contrast photographs, which reduced the sample to a near silhouette, did produce further decrements in performance, but this form of alteration also eliminates many of the internal relationships that are presumably used to make such transpositions (Davenport and Rogers 1971; Gardner and Gardner 1971; Hayes and Nissen 1971). The home-reared chimpanzee Vicki was also systematically tested on concept discrimination using pictures, and she could discriminate animate from inanimate, male human from female human, and children from adult, as well as some geometrical figures. The sign-language trained apes can also recognize pictures in books and produce the sign language name of the concept in question. Lehr also tested a rhesus and a *Cebus* monkey with pictures and trained them to discriminate pictures of insects and flowers (Lehr 1967). These monkeys could generalize the concepts to pictures they had not seen before and could discriminate abstract flower designs from other patterns. These experiments demonstrate conceptualization in monkeys and apes, and Rogers and Davenport argue that their delayed cross-modal matching results imply a symbolic process. However, sensory concepts still have an iconic component, and *symbol* has been so variously defined that it might be better to hold the term in abeyance until other conceptual types are considered.

Many of the concepts humans use as a matter of course are perceptually *disjunctive,* in that there are no internal perceptual relations shared by all possible exemplars. What have been called *instrumental* concepts often have this property (Bruner et al. 1956). For example, *hammer,* which the dictionary defines as a "tool for pounding," embraces the disjunctive class of objects that can serve that purpose. A trip to the hardware store will reveal that there are many kinds of hammers, some of them with rubber heads, and many kinds of objects can be used as hammers if nothing else is available. Consequently, the concept *hammer* may be quite different psychologically from the morphological concepts most commonly used in studies of primate cognition. In instrumental concepts, the equivalence among exemplars is not based on the abstraction of perceptual commonalities but on the abstraction of the commonalities in the consequences that follow from their instrumental use. Instrumental concepts are classifications based on cause and effect relationships. Perceptual recognition criteria are normally bound to instrumental concepts, just as most people can recognize a "normal" hammer and perhaps even look for a particular hammer when confronted with a need to pound, but the creative aspect of tool use requires the ability to selectively deactivate such normal perceptual recognition criteria and concentrate on the functional relationships. Even a cursory examination of the human world reveals that many everyday concepts are in fact more relational and perceptually disjunctive than a hammer, in that the perceptual recognition criteria are subordinated to purely relational judgments. A *strike* in baseball, for example, can be a pitched ball that is swung at but missed, a ball that should have been swung at but was not, or a ball that was hit but went foul, depending upon whether it is caught by the catcher and whether there are already two strikes by the same batter. Experience with still photographs of behaviors that have been called strikes would not be sufficient for abstracting the conceptual definition, even though this procedure is sufficient for teaching object concepts, as experiments with animals have shown. The rules of baseball are presumably within the intellectual competence of all normal human beings, however incomprehensible cricket may be, and it is not a particularly complex example. All human social organization requires systems of perceptually disjunctive relational concepts, the sophistication of which is best revealed by reflecting how they might be taught to animals by multiple presentation of exemplars. Anthropologists have shown, for example, that all human societies organize basic social relationships in reference to systems of kinship reckoning in which disjunctive relational concepts play an essential part (Chambers and Tavuchis 1976; Tyler 1969). No amount of contact with exemplars of uncles, for example, would allow one to abstract the perceptual recognition criteria, for the concept is purely relational. That is to say, the perceptual judgment of who is an uncle and who is not is dependent upon conceptual relationships to which the stimulus features of particular exemplars are irrelevant. Support for this is provided by studies of kin term acquisition in children, where actual family composition is an inconsistent predictor of whether the term is known. It can be confidently asserted that human beings are the only known animal whose social activities are dependent upon categorizations of perceptual events in terms of recognition criteria that are themselves conceptual, and it is difficult to discover to what extent other primates share this capacity. Without language, it is very hard to ask an ape what you want to know, and the hiatus in the comparative record may simply reflect an inability to pose the question in terms an animal can understand. Nonetheless, it can be demonstrated that animals can learn purely relational and disjunctive concepts, even if it is still an open question as to whether animals have systems of relational concepts organized in quasialgebraic form.

Relational concepts are also demonstrated in animals by testing procedures that eliminate alternative forms of explanation. The usual procedure is to randomize all aspects of the presentation except the relationship which is being taught. If the particular perceptual properties of a given test stimulus are irrelevant to the solution of the problem, but the animal nonetheless chooses correctly, it is concluded that a relational concept has been formed. This ability to abstract a concept from a number of rewarded instances and to generalize it to new instances has been extensively studied in primates under the name of *discrimination learning sets* (Miles 1965; Kintz et al. 1969; Rumbaugh 1970, 1975; Warren 1965; Brown et al. 1959).[2] Harlow has characterized these tests as providing evidence for "learning to learn," and it is indicated by the fact that the number of exposures or trials that the animal takes to reach criterion performance is reduced as more stimulus pairs are presented, indicating that it has acquired a set that is not restricted to specific stimulus pairs. A monkey skilled in discrimination learning tasks eventually learns a rule that it applies to any pair of stimuli in a testing apparatus: if the response to one is rewarded, then stay with that one; if not rewarded, then next time choose the other member of the pair. Psychologists call this a win–stay/lose–shift strategy.

In discrimination learning tasks where the stimulus objects are purposely made as variable as possible from one problem to the next, the relationship between the reward and nonreward is the only concept that can be abstracted from the situation. Nonetheless, these tests can be used to assess a monkey's retention for specific stimulus pairs, and rhesus monkeys have successfully remembered the reward–values of seventy-two different stimulus pairs, even when tested 30 days after the last experimental session or 210 days after the original learning to criterion performance (Strong 1959). If stimulus pairs are presented that exemplify a specific concept, then multiple discrimination learning problems can provide unequivocal evidence for concept acquisition, since the concept to be learned is chosen by the experimenter and conveyed to the animal by the manipulation of reward and nonreward. If the animal's performance shows improvement with increasing experience of exemplars and nonexemplars of the concept to be learned, then eventually it should be able to perform perfectly with pairs of stimuli it has never seen before, as with Lehr's flower and insect pictures. Moreover, it should also be possible for primates to learn relational concepts in which the specific sensory properties of the test stimuli are irrelevant to the choice of the correct object. Both chimpanzees and macaque monkeys have kept pace with psychologists' ingenuity in designing learning set problems.

In one experiment testing relational concepts, which relied upon the proven ability of rhesus monkeys to remember particular objects, the monkey had to displace an object on each trial to obtain a reward from a food well underneath. Four different objects were used, and they were always presented one at a time in the same sequential order (Massar and Davis 1959). In a subsequent experimental session, two of the objects were presented simultaneously on a particular trial, and the monkeys had to choose the object that was in the correct sequential order for that trial. These monkeys performed at better than chance but they did not reach the same level of proficiency as with other discrimination problems. In another test, the oddity problem, the animal is presented with three stimuli, one of which is odd when measured on a particular dimension of contrast to the other two (Bernstein 1961; Young and Harlow 1943*a*, 1943*b*). The odd stimulus is always positive, and the position of the objects is randomly varied between trials to negate any constancy of location which could serve as a cue. Harlow demonstrated that rhesus monkeys can form oddity learning sets, in which they learn to choose the odd member of a triad. Bernstein carried these tests to a higher level of sophistication by showing that primates can learn what are called *dimension-abstracted* oddity problems, in which the dimension of contrast varies from one problem to the next. For example, in a set of five objects, four of which are similar in size but all five are the same color, then the dimension of oddity is size and the odd object is the one that is smaller than the others. In another problem, the set of five objects may have five objects of dissimilar shape, four of which are the same color. In this case, color is the dimension of contrast for oddity, and the oddly colored object is correct. Both great apes and macaque monkeys have been tested with up to seventy concurrent dimension-abstracted oddity problems. The apes learn faster than monkeys, but both can learn them. In matching-to-sample, in some ways the converse of oddity, the animal has to choose the member of a pair of objects that matches a sample object. Learning sets can be formed to matching-to-sample tasks, and primates can also match on *dimensions* of similarity, such as color or shape, which are independent of prior experience with particular stimulus arrays (Nissen et al. 1948, 1949). A further aspect of monkey cognition is the ability to use arbitrarily learned cues to reverse the reward value of a stimulus. For example, in *conditional* discrimination tasks, it is possible to teach monkeys to choose one member of a pair of objects if one is presented on a tray of one color and the other member is presented with a second color (Barge and Thomas 1969; Nissen 1951; French 1965; Riopelle and Copelan 1954). Dewson has recently combined the conditional discrimination technique with other procedures to produce the most complex discrimination problem yet taught to monkeys, in this case, *Macaca fascicularis*. The animals are first taught that a particular sound matches a particular colored light. After this arbitrary equivalence is established, the animals must listen for an acoustic stimulus, remember it, compare it mentally to a second stimulus occurring later – after an interval from milliseconds to seconds – and then order their response on the basis of the order of the acoustic stimuli (Dewson 1977; Cowey and Weiskrantz 1976). The subjects convey the order of the acoustic stimuli by pressing the colored lights in the order corresponding to the order of the sounds they heard. Hence this test is sequential and involves different sensory modalities. By extending the time before the colored lights are available for responding, Dewson has been able to study a monkey analog of auditory short-term memory, a phenomenon hitherto restricted to study in man. Interestingly, the ability is impaired by a lesion of the auditory association cortex, area twenty-two, which corresponds to the receptive speech area in man—but only if the monkey's lesion is in the left hemisphere (with five monkeys tested). Further evidence for short-term memory in nonhuman primates has also been shown for object concepts. Rumbaugh and his colleagues successfully trained young chimpanzees and orangutans to retrieve objects that they could observe being hidden, and Menzel has performed similar experiments with chimpanzees in a semi-naturalistic setting (Menzel 1971, 1973). The chimpanzees were held in a small enclosure but could watch while the experimenter hid food or toys in various places. When released, they retrieved the objects by going directly to the hiding places.

All of these experiments have emphasized conjunctive concepts with a specifiable common feature. The most persuasive evidence for disjunctive conceptualization in animals is provided by the language analog experiments. Premack has devised an experimental design that uses plastic figures with metal backs that can be attached to a magnetized board (1976; Premack and Premack 1972). Each physically distinct plastic piece represents a meaningful unit, analogous to an English word or name, such as *banana, apple, Mary, red,* and so forth. The names for objects are originally taught by pairing the object with its plastic name, henceforth called a *tag.* The chimpanzee must place the tag on the board before receiving the object. In a later stage, the chimpanzee had to place both the tag of the object it wanted and the tag for the name of the tutor who would give the object. Moreover, the animal had to observe a standardized order in placing tags on the board. In a subsequent stage, the tag for the recipient of the object was introduced, forcing the chimpanzee to produce strings of tags such as "Mary give apple Sarah." Tags introduced in this way denote concepts and not single stimuli, as can be shown by their transfer to novel exemplars. Moreover, it is possible to test the hypothetical concept by using the matching-to-sample test. The tag for *apple,* for example, is used as the sample stimulus, and the chimpanzee must choose between pairs of objects and also pairs of other words, such as red/green or round/square. In these situations, the chimpanzee chooses the right object and also the tags that are most descriptive of the apples it has seen, namely red and round. This choice is inexplicable if the animal were responding to the tag for apple as an object in its own right, for the word for apple is a blue triangle.

By placing tags and objects together in "hybrid" combinations, it has been possible to teach a number of other dependencies, analogous to conditional discrimination tasks. The chimpanzee was taught, for example, to place a particular irregular blue polygon, which can be designated IBP, between an object and its tag. For example, if presented with an apple and a blue triangle, Sarah was to place IBP between the two. Adopting a notation that places *obj* after objects and *tag* after tags, Sarah's correct solution to this task would be the string "APPLE tag + IBP tag + APPLE obj." Similarly, a second blue polygon, shaped differently from the first, was introduced as a contingency for those pairs that were not related as tag and concept. This second tag, here designated IBP2, was to be placed before IBP in anomalous strings. Thus if presented with the string "APPLE tag + BANANA obj," Sarah's correct solution would be the string "APPLE tag + IBP2 tag + IBP tag + BANANA obj." This procedure could be transferred to pairs of concepts and tags that were not part of the training, indicating that it was encoded on a relational level and not tied to particular exemplars. The strings that were correct solutions to the IBP task could also be used as conditional cues in a yes/no test. Having been taught tags for *yes* and *no,* in the context of correctly and incorrectly formed strings, Sarah could use strings containing IBP and IBP2 to make the correct choice between the tags *yes* and *no.* Thus the correct answer to "APPLE tag + IBP tag + BANANA obj" is "NO tag"; whereas "APPLE tag + IBP2 tag + IBP tag + BANANA obj" is "YES tag." Again, these relations are conceptual rather than stimulus-bound because the string "FIG tag + IBP tag + FIG obj" was sufficient to introduce the tag for fig, a symbol she had not seen before. Subsequently, when presented with the array of tags *fig, Mary, give, Sarah* and a few others, Sarah produced the string, "Mary give fig Sarah." A similar procedure was used to teach the tag *color-of,* and then the COLOR-OF tag was subsequently used to teach novel color words. The conceptual nature of the novel tags is also indicated by transfer tests, where BROWN tag, for example, was used correctly by the animal to choose among four colored disks, only one of which was brown. The tags introduced in this way also function productively within the context of the other conditional rules established previously. Sarah could respond correctly when confronted with the string "SARAH tag + INSERT tag + BROWN tag [disk in] + RED tag + DISH tag."

Sarah's tag-using performances provide particularly good evidence for conceptualization and transfer in apes, but it is difficult to specify exactly what the animal has learned. The data on naturalistic tool use indicate that the motor skills employed in these tests — replacing one object with another, inserting an object into a container, and putting down objects in sequence — are already part of the chimpanzee's normal skill repertory. The conceptual repertory of object types and different varieties of fruit can also be considered normal in chimpanzees and need not be explicitly taught. However, the interrelationship of concept and motor act, since it is created through an arbitrarily chosen intermediate cue, must be acquired in the course of the training. What is most striking about Sarah's performance is her willingness to form equivalence classes in which the dissimilar morphology of tag and object is ignored. This indicates that chimpanzees can operate with disjunctive conceptual classes when the exemplars are instrumental to the same effect. Moreover, these disjunctive concepts have contextually sensitive spheres of applicability. Sarah recognizes the equivalence of an apple and a blue triangle when they are used as a cue in a match-to-sample problem, but she does not make the error of trying to eat the plastic chip. Consequently, Sarah's performance requires the postulation not only of distinct concepts, both disjunctive and conjunctive, but also of contextualized conditions that specify their sphere of applicability. The question of whether the Premack experiments are a good language analog is far less important than what the experiments actually show about chimpanzee cognition. They indicate that chimpanzees not only can abstract constant relationships on the basis of experience and encode this information in a conceptual network, but they can also encode the contextual cues that activate a particular conceptual schema, such as the situations in which a plastic chip can substitute for an apple.

Cognitive tests of nonhuman primates are important to an evolutionary perspective because they indicate that primates possess sophisticated cognitive abilities which do not necessarily have directly observable correlates in naturalistic situations. Moreover, they indicate commonalities of conceptual information processing, at least among humans, anthropoid apes, and macaque monkeys: all of these species have the ability to form concepts after repeated exposures to pairs of exemplar and nonexemplar; to use the concepts thus formed to discriminate new instances; to alter the dimensions of contrast and similarity by which exemplars are being discriminated; to retain concepts for long periods of time; to retain multiple concepts simultaneously; to retain particular concept exemplars in memory; to form relational concepts that are independent of particular stimulus properties; and to learn arbitrary signals that can cue the relevant dimension of discrimination.

Furthermore, extensive experimentation has shown that the integrity of neocortical mechanisms is required for the conceptual assessment of the external factors leading to reward and nonreward. As is well known, primate evolution has been characterized by a progressive elaboration of neocortical mechanisms, and learning set performance is itself phylogenetically progressive, showing quantitative improvement from primitive mammals, to monkeys, to apes, to man. Yet in none of the learning set tests just discussed is the *behavioral* component of the task much more complex than pressing a button or retrieving a peanut. A purely morphological description, based upon these behavior patterns, without any attempt to characterize the controlling information, would be a serious distortion of the animal's behavior. These facts suggest that these sophisticated abilities to process conceptual information coexist, in the case of monkeys at least, with a largely innate behavioral repertory. Consequently, *the progressive evolution of primate cognition does not depend upon the replacement of innate behavior by learned behavior but by the selection and control of innate behavior by conceptually stored information.* The building blocks of behavior are phylogenetically conservative, but the progressive development of cognition allows the integration of very old motor components with information acquired in the course of experience. This approach resolves some of the contradictions in the chimpanzee literature, such as the role of experience in the development of termiting behavior and Schiller's observation that all of the motor components of termiting, such as inserting sticks into holes and tearing the leaves off branches, appear in captive chimpanzees who have never observed termiting (1952, 1957). However, in a specific test of the relative abilities of wild-born and cage-reared chimpanzees to solve a problem using a tool, the cage-reared chimpanzees were still largely incompetent, even though the motor patterns were present in both groups (Birch 1945; Menzel et al. 1970). If conceptualization is regarded as an interface network between the innate motor repertory and the innate desires of the animal – a network in which the patterns of connectivity are created through experience – then it is quite possible for the development of termiting to require both conceptual experience with other termiting chimps and the maturation of innate behavior patterns.

The evidence on primate cognition also argues against purely behavioral definitions of tool use, which equate tools with material objects. Such definitions are useful to the extent that they call attention to similarities across species, but psychologically they make an error exactly analogous to describing, say, a conditional, sequential cross-modal discrimination task as a form of bar pressing. Primate tool use is not a kind of behavior but a modality of action that uses *peripheral effectors to create external effects that have been predicted from conceptually stored information.* The selection and control of motor patterns by conceptually stored information at the highest level of motor integration can be termed the *instrumental modality of action,* and probably all species of primates are capable of such control to some degree, even if their motor patterns are completely innate.

The special-sensory cortical areas and associated intrinsic cortex are involved in processing the context-free regularities in sensory information (Rumbaugh and Gill 1973).[3] These regions expand in the course of primate evolution as increased instrumental control would demand. It is important to note that the instrumental modality of action subsumes the distinction between innate and learned behavior on one hand and between affective and nonaffective action on the other. Both learned and innate behaviors can be controlled by the assessment of the external consequences of action, and the external consequences of action are themselves constantly monitored by the affective modality for their emotional implications. The instrumental modality also differs in an important way from the instinct/learning interlockings of Lorenz. Learned and innate components are not distinct slots within an instinctive chain but participate in a common network that interrelates all behaviors with similar external effects.

This approach to the primate nervous sytem, which rejects the distinction between learned and innate behaviors as of fundamental importance and substitutes functional divisions based on the assessment of the internal and external consequences of action, makes better neuropsychological sense than the higher and lower centers of the Victorian brain, as Pribram has argued. Furthermore, it suggests that both modalities of action, the affective and the instrumental, have access to both learned and innate motor programs for the implementation of their respective goals. The affective modality has access to instrumentality with its context-free sensory processing, conceptual inference, and learned skill repertory. The instrumental modality in turn can call upon the conceptual encoding of the internal consequences of external acts to carry out goal-directed action on others, and it can also control the innate behavioral components of affective action through increased voluntary motor control of visceroautonomic processes. Thus conceptual encoding involves the interrelationship of the internal and external consequences of action.

The concept of an instrumental modality of action has important consequences for the study of primate evolution because it implies that goal-directed, instrumental action should be far more widespread than behavioral definitions of tool use would suggest and that neocortical monitoring should be common, even in animals with few learned behavior patterns. There is physiological evidence that this is the case (Talmadge-Riggs et al. 1972; Newman and Symmes 1974; Wollberg and Newman 1972; Ploog 1967). Even though the vocalizations of squirrel monkeys *(Saimiri)* are clearly innate, in that they develop normally in deafened animals, and are subcortically mediated, as shown by electrostimulation experiments, which can elicit them from the limbic system but not from the cortical larynx area, recordings from single neurons in the superior temporal (auditory) cortex nonetheless show that tape-recorded vocalization is neocortically monitored. In rhesus monkeys, the vocalizations are also innate and elicitable from subcortical limbic areas, but the animals increase the volume of their conditioned calls in the presence of external noise, suggesting that these vocalizations are also being at least partly controlled by sensory evaluation, although it is not known whether this is cortical (Robinson 1967; Sinnott et al. 1973; Sutton et al. 1974).

The concept of an instrumental modality also allows for differences among primates in the extent to which specific effector systems can operate under instrumental control. Since the instrumental modality implies neocortical sensory and motor mechanisms, the extent of neocortical control can be used as a measure of the extent to which an effector system can be incorporated into skillful programs. Myers has argued, for example, that the motor patterns of facial expression are less neocorticalized in rhesus monkeys than in man because lesions of the motor face area in the former do not produce weakness in facial expression on the side opposite the lesion. Similarly, it is well known that the vocal tract of nonhuman primates is under little volitional control because the behavior produced is innate and difficult to condition operantly. This suggests that the instrumental use of affective programs is a more phylogenetically progressive phenomenon in Old World anthropoids than either the affective use of instrumental programs or the conceptual contextualization of affective programs.

The integration of preexisting systems into the instrumental modality is also observed in many other aspects of primate behavior that do not conveniently fit the definition of a tool as an unattached material object. Instrumental behavior is frequently directed to the physical substrate, as when a chimpanzee bends down tree branches to make a nest or when a spider monkey bends a branch across a gap between two limbs to make a bridge for a baby. Moreover, the body itself is usually the agent in instrumental action and participates in the instrumental mode. This observation does not overextend the definition of tool use but recognizes the important biological fact that instrumentality uses the skeletal musculature and the related programs of posture and movement as its major avenue of operation. In primates, the affective mechanisms are to some extent integrated into the instrumental modality, and other individuals are frequently used in the manipulation of cause and effect relationships. A good example is provided by the phenomenon of *agonistic buffering*, especially elaborated in *Macaca sylvanus,* in which adult males retrieve infants and carry them over to other adult males, to whom they are presented (Deag and Crook 1971). This phenomenon, also sporadically occurring in *Macaca fuscata,* is not so much a form of male care giving as it is the use of infants in adult male status transactions. Since it is possible that infants have an inhibitory effect on aggression by evoking the incompatible maternal motivations, the presentation of an infant may allow greater social proximity between the males. The skillful manipulation of these biological contingencies, however, is unlikely to be completely innate but probably requires a conceptual understanding of social facts. A related example is the "tripartite threat," well described for *Papio hamadryas,* in which a low-ranking individual, monkey C, stations itself between two higher ranking monkeys, A and B. Monkey C then threatens B while keeping its back to A. Since A outranks both B and C, the victim of the threat cannot counterattack without apparently threatening A as well (Kummer 1967). Since strategies of this kind require noninnate information about the social ranks and spatial relationships that exist at particular times, it is reasonable to infer conceptual control of the affective act. However, it is certainly pertinent to the concept of instrumental/ affective interaction that in both protected threat and in agonistic buffering, the skillful manipulation involves skeletal muscle systems already known to be volitionally controlled in monkeys — infant retrieval and spatial positioning of locomotion and posture — and not the volitional control of the affective expression itself.[4] This latter ability — dissimulation in affective expression — may well be a hominid skill (Ekman and Friesen 1969;

Ekman et al. 1976). In both examples, however, the behavior is affectively motivated, conceptually informed, instrumentally controlled, and integrated with innate fixed-action patterns, and no one of these predications is more true than any other. One aspect of advanced primate behavior is that it is all of these things at once.

The progressive sophistication of conceptual processing, revealed by phyletic trends in learning set performance, mirror image concepts, and sign language proficiency, does not occur in isolation but is paralleled by a progressive alteration of primate motor skills, especially in the capacity to manipulate material objects. This trend is well illustrated by Parker's comparative study of the way in which zoo-housed primates handle a rope (tables 4.1 and 4.2) (1974). Parker distinguished between *primary* actions, in which the rope is acted upon, and *secondary* actions in which the rope is used

TABLE 4.1   *Primate actions involving a rope*

I. Primary Actions

    A. Nongrasping contact

        1. Touching without moving the rope (1)*
        2. Supporting against gravity without movement (2)
        3. Raking toward the body with curved fingers (1)
        4. Picking at the surface with finger nail (1)
        5. Pushing away from the body (3)
        6. Moving the BP along the surface (5)
        7. Resting on, without moving it (5)
        8. Jumping onto the rope (1)
        9. Throwing it off the back by bucking the back and/or hunching the shoulders (2)
      10. Flipping upward with rapid finger action, palm upward, little or no arm movement (1)
      11. Drawing into the mouth using only the lips (1)
      12. Biting with the teeth (1)
      13. Licking with the tongue (1)
      14. Striking with the BP, a high intensity action (1)
      15. Patting or tapping on, a low intensity action (2)
      16. Wadding up into a compact ball or mass (2)
      17. One hand orbiting the other, with the rope over or around the hands (1)
      18. Circular swirling motion with BP on the floor (1)
      19. Tossing from open hand to open hand (1)
      20. Walking with the rope wrapped around both ankles or with legs in the space between two separated strands (1)
      21. Two BPs pushing in opposition to each other (1)

Table 4.1 (cont.)

B. Hand grasp plus additional action

22. Picking up off the floor (2)
23. Holding without movement (6)
24. Holding lightly, allowing it to unwind after being twisted (1)
25. Pulling in opposition to the mooring (16)
26. Jerking, a rapid alternate push-pull (2)
27. Throwing, always with a loss of contact at the end of the action (4)
28. Twisting by means of forearm rotation (1)
29. Wrapping around the BP by turning that same BP (1)
30. Mixing up, like tossing a salad (1)
31. Walking about while carrying (1)
32. Pulling by locomoting away from the point of attachment until slack is removed (4)
33. Shaking by a rapid alternation of the direction of movement (3)
34. Waving in a slow alternating or circular motion (3)
35. Rolling, spinning, or rocking while holding (4)
36. Two or more BPs pulling in opposition to each other (1)
37. Pulling facilitated by applying another BP to the cage bars to push or pull in opposition to the primary action (3)
38. One BP pulling it taut, another plucking or pushing taut part (3)
39. Unwinding one or two strands while holding other two or one (1)
40. Pulling away from the body, hand between the bars outside the cage (1)

II. Secondary Actions

C. With respect to another body part

41. Draping over another BP, allowed to hang down by gravity (18)
42. Pressing against another BP (2)
43. Holding taut with one hand, turning body to wind rope around body (1)
44. Inverse draping, around BP with ends held up against gravity (3)
45. Sliding across another BP by pulling (3)
46. Pushing into the mouth with another BP (1)
47. Patting rope against another BP (6)
48. Striking another BP with it (1)
49. Rubbing on another BP (3)
50. Forming a loose loop about a BP (1)
51. Wrapping tightly about another BP (11)
52. Forming a circle around the body on the floor (1) and sitting, standing, or lying in the circle (3)

Table 4.1 (cont.)

D. With respect to an object

    53. Rubbing rope on itself (1)
    54. Forming a loop with nothing inside it (5)
    55. Wrapping rope around itself (1)
    56. Drape over object (1)
    57. Squeeze rope and object in same hand (1)
    58. Two BPs pulling in opposition with loop around an object (1)
    59. Pulling in between two cage bars (1)
    60. Pushing out between two cage bars (1)
    61. Pushing into a corner of the cage (1)
    62. Pressing against an object (3)
    63. Rolling rope on floor back and forth with palm (1)
    64. Patting on rope held against an object (1)
    65. Striking an object with the rope (2)
    66. Rubbing it on an object (2)
    67. Smearing feces with it (1)
    68. Wadding up under or between objects (1)
    69. Forming a loop around an object (1)

SOURCE: C. E. Parker, "The antecedents of man the manipulator," *J. Human Evolution* 3 (1974): 493–500. With permission.

*Numerals in parentheses indicate the number of specific actions subsumed under a generic action; and BP stands for body part.

in reference to something else, including one's own body. It can be seen that the great apes (table 4.2) use more secondary manipulations than do monkeys, which is in accord with Gallup's findings on their respective abilities to incorporate their own self-images. The ability to use a material object as an effector requires the prior ability to maintain a tonic holding pattern between a material object and a body part (termed a tonic GRIP function) and to incorporate this extension of the body into a sensory/motor map of this new composite structure. From a psychological perspective, it is this ability to incorporate objects held in the hand into an external map which is the significant aspect of material object use, not the fact of the materiality of the object itself. Consequently when macaque monkeys take their own tails into their two hands and thread the tail through a hole in the fence to retrieve a peanut (figure 4.1), this is the genuine instrumental use of an effector, regardless of whether the tail is part of the body: the guidance is performed by the hand under the tactile feedback from the tail, as the monkey often performed this skill with its head turned away (Erwin 1974).[5]

TABLE 4.2   *Species in rank order for mean percentage of behavior classified as secondary*

Species	Percent
Orangutan	35.59
Chimpanzee	28.73
Gorilla	12.81
Capuchin	6.25
Macaque	4.85
Gibbon	3.54
Lemur	2.84
Langur	2.05
Guenon	0.00
Spider Monkey	0.00

SOURCE: C. E. Parker, "The antecedents of man the manipulator," *J. Human Evolution* 3 (1974): 493–500. With permission.

A comparison of tool use in nonhuman primates reveals a phylogenetic progression in several primate lineages in the ability to incorporate different object manipulation programs into the instrumental modality. Parker's data suggest an elaboration of this capacity in the lineages leading to the anthropoid apes, the macaques, and baboons among the Old World monkeys, and the capuchin *(Cebus)* monkeys in the New World. These same species have also generally shown the greatest manipulative skill in naturalistic situations and in laboratory experiments (Beck 1974, 1973; Bolwig 1963, 1961; Bertrand 1967; Chiang 1967; Hamilton et al. 1975; Lawick-Goodall et al. 1973; Eaton 1972; Vevers and Weiner 1963; Kawai 1975).

Evidence for tool use in naturalistic settings has been found for chimpanzees throughout their range, with some regional variations (Struhsaker and Hunkeler 1971; Teleki 1974; Nishida 1973; McGrew 1974; Jones and Sabater Pi 1969; Sabater Pi 1974). Goodall had observed chimpanzees in Tanzania probing for termites with twigs and grass stalks, and Suzuki has reported evidence of chimpanzees using termiting sticks elsewhere in Tanzania. Termite mounds with probing sticks in place have also been observed in chimpanzee habitats on the other side of the African continent, in Rio Muni. In Cameroon there is evidence that chimpanzees use hammer stones to smash nuts, and hitting with sticks has also been widely reported from chimpanzee agonistic displays. Beck has reviewed a now extensive literature on observations of primate tool use, and many species of monkeys and all apes have been reported to throw and drop objects as part of defensive behavior (1975). All apes are also capable of using spongelike

FIGURE 4.1 *Macaca fascicularis* using its tail as a tool to retrieve a peanut: *A*, hand puts tail through fence; *B*, tail sweeps peanut. *(Photos by the author.)*

objects, such as cloth or wadded leaves, to soak up moisture. All of the anthropoid ape species have been observed to hammer objects, drape material over their bodies, hit with sticks, insert or probe with sticks, use sticks as extensions of the arm in reaching, balance sticks to facilitate climbing, and use sticks as levers. Macaque, baboon, and *Cebus* monkeys also use sticks as extensions of the arm in reaching, applying leverage, inserting, and hitting. Hammering also occurs in these monkey species, as does cleaning with a spongelike object.

However, captive monkeys have not reached the levels of object-use that are easily demonstrated in captive anthropoid apes. The stacking of boxes to reach food suspended from the roof is well known from Kohler's experiments, as is the performance of the chimpanzee Sultan, who could put short sticks together to make a long one (1959). In the combination of behavioral acts into a complex, goal-directed sequence, anthropoid apes show a clear superiority to monkeys. Döhl taught a chimpanzee a complex Chinese box type of puzzle (1966). In this experiment there were two series of five boxes each, both presented to the animal simultaneously so as to make an array of ten boxes. The goal box of one series is empty whereas the other contains a reward. To open the goal box, the animal must first open the goal-minus-one box and get the key. However the key to the goal-minus-one box is in goal-minus-two box, and so on, out to the goal-minus-four box of each series. The keys to the outermost boxes of each series are hung in a binary choice apparatus, and the animal must choose the key that leads to the reward through the five intermediate boxes. The wrong key, of course, leads to an empty goal box. The key contained in each box is visible through the transparent cover, but to choose the first key correctly, the animal must reason through the series of boxes to the goal box containing the reward. In other experiments, great apes have been taught as many as fourteen separate manipulations in opening a single box, including a sliding bolt, a padlock and key, a hook and eye, removable pins and eyes, and strings tied in bows (Rensch and Döhl 1967). Home-reared apes also learn a number of complex manipulations, often without direct tutoring. The chimpanzee Lucy would untie her owner's shoes and put them on herself and would often unbutton shirts (Temerlin 1975). Lucy will also turn on the kitchen tap and fill a glass with water, plug in electrical appliances and turn them on, and operate cigarette lighters to blow out the flame. Recently, Rumbaugh and colleagues have shown that chimpanzees can use rule-governed strings of tags to request a tool from another chimpanzee to solve a particular manipulatory task (Savage-Rumbaugh et al., forthcoming).

The tool-using performances of tutored and home-reared apes show a comparable variety of domestic skills, which suggests that apes have a level of conceptual processing sufficient for grasping many of the causal relationships needed to operate within the domestic sphere of an existing complex technology (though not in a traditional technology!). However, the Russian primatologists suggested long ago that the ability to analyze causal relationships, which they termed *analytic* intelligence, was more advanced than the ability to implement such causal understanding as part of a constructional process, which they called *synthetic* intelligence (Ladygina-Kots and Dembrovskii 1969). Even apes highly skilled in the operation of domestic tools and machines do not show a comparable level of sophistication in the synthetic aspects of object use. The stacking of boxes, the breaking of a stone with a hammer to make a flake (with help), the stripping of leaves off sticks to make probes, the joining of short sticks to make a long one, and the placing of tree limbs to make ladders represent the current limits of ape constructional ability with objects (Menzel 1972; Wright 1972; Beck 1975; Kohler 1959). This repertoire is impressive in the context of other primates, but the limitations of ape synthetic abilities are most dramatically conveyed by comparing them to normal human levels of object construction.

For example, there is nothing profoundly technical in making the kite shown in figure 4.2, and the procedures have been described in the book *Kites to Make and Fly* in a series designed for children in the ten to fourteen year age group (anonymous 1976). If the major operations are abstracted from the text, they can be listed as follows:

1. measure sticks
2. cut sticks to length
3. notch the ends of the sticks
4. test for balance
5. notch the center of the sticks

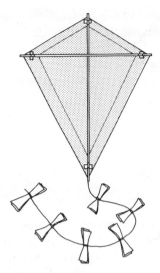

FIGURE 4.2 A kite made from instructions in a book. *(Sketch commissioned by the author.)*

6. join the sticks with glue
7. reinforce the join with knots
8. attach tie strings to the ends of the sticks
9. attach the frame string
10. tighten the frame string
11. test the frame for symmetry
12. measure paper against frame
13. cut paper to size
14. attach paper to frame string
15. attach "bridle" string to frame
16. make a number of paper tail feathers
17. attach feathers to tail string
18. attach tail to kite

These operations are well within the technical competence of normal, adult humans, and few readers of the present book would have difficulty in performing any of them, even if they had little practical experience with carpentry. However, the eighteen operations are themselves complex skills with a great number of constituents. Consider the third item, notching the ends of the sticks to keep the knots from slipping off, which is simple enough in principle. What is actually involved? Notching can also be written as a list of operations:

1. locate the top notch at the left end of the stick
2. locate the right arm of that notch

3. align the knife blade in three dimensions
4. apply pressure to knife until cut is made
5. locate the left arm of that notch
6. repeat (3) and (4) until the notch appears
7. rotate the stick on its long axis to locate a new notch
8. repeat (2) to (6)
9. rotate the stick end over end to make the former right end the new left end
10. repeat (1) to (8)
11. put down the stick
12. locate the second stick
13. repeat (1) to (10)

This sequence is also an easy one for humans to grasp, but there are anomalies to the description that become glaring when viewed in an evolutionary context. It should be apparent that an unnotched stick has no notches, and this creates problems in the direction of motor activity. The knife blade must be directed to the place where the notch will be after the notching operation is complete. Real perceptual content is no help here, any more than it is with relational concepts. No amount of perusal of the unnotched stick will reveal the location of the notches. The book recommends that the notches be placed fifteen millimeters from the end of the stick, but this instruction will be no help to a protohominid because it requires language, as is also indicated by the fact that loss of the ability to make arithmetical calculations is a common consequence of damage to the speech hemisphere in modern man. The easiest way to make the notches is to imagine the stick with notches in it and direct the knife to the real location of the imaginary notches. Archaeologists who have been simulating extinct stone tool technologies, using flint and hammer stones, have come to a similar conclusion. These workers have all postulated the ability to impose a mental template on unworked flint (Bordes 1971). Holloway, too, in reflecting on the technological differences between hominids and apes, has argued for the capacity of hominids to impose arbitrary form upon shapeless material (1969). This hypothetical ability is certainly in accord with common human experience, which the reader can verify for himself by closing his eyes and imagining the kite in figure 4.2 to be red with white spots. Physiologically, there is a well-known precedent for internal imagery even in nonhuman primates, as the neurons of the visual cortex of monkeys are as active during dreaming sleep as during wakefulness, even though the animal is more refractory to external stimulation than in any state except coma.[6]

There also is evidence that monkeys can operate with encodings of images during wakefulness that remain stable for as long as a week. Davis, working with three species of macaque monkeys, demonstrated that the monkeys could remember the precise size of a geometrical figure. The monkeys were trained on a match-to-sample test, in which a sample stimulus was presented and then removed (Davis 1971). The subjects were then given simultaneous stimuli, one of which was the sample, and they had to choose the one they had seen before. In the next stage they were presented with the test stimuli, all circles differing only in size, without exposure to the sample, and they chose the correct stimulus from memory. Even if tested a week later, the monkeys could still discriminate the correct stimulus from others close to it in size. This experiment indicates that monkeys can use size information stored in memory to discriminate perceptual content. However, there is still a subtle difference between the performance of Davis's macaques and the act of imagining notches on a stick. The monkeys could direct motor behavior to the external region that *matched* a mentally encoded image, but there is very little indication that nonhuman primates can direct behavior to an external region that exemplifies a mental image when there is *no perceptual match* in the external world. One way of formulating this difference is to say that humans can (1) map a simulated image on to a perceptual image and (2) direct motor behavior to points in the external world that correspond to points on the imaginary image. A map, after all, is not a facsimile of something else but an orderly transformation, such that the relationships between points on a map also obtain between points on the referent if the reverse transformation is applied. In the mapping of one image on to another, the perceptual world is itself to some extent rendered imaginary, as perceptual judgments are being made on the basis of purely mental criteria. In other words, hominids not only utilize conceptually encoded recognition criteria, like the strike in baseball, but frequently have a need to literally see things that are not there. This ability should be distinguished from the prediction of external effects already discussed in reference to conceptual information processing. An ape can presumably predict certain kinds of consequences, not yet observable, when it plans on the basis of conceptual information. Yet even if one can predict that notches will appear on a stick if it is subjected to certain operations, one still needs a way of controlling motor activity in reference to still nonexistent perceptual features to perform the operations themselves. To validate a conceptual prediction, in contrast, one need only wait and see.

This theoretical approach postulates both conceptual and behavioral continuity between humans and apes but indicates a discontinuity in the capacity of internal imagery to control the motor acts during constructional tasks. It predicts that apes will be able to construct any object that requires only real-world perceptual information: a real box placed on top of another real box or the removal of real leaves from a real stick. An important test case would be the careful examination of ape nest building to see if a superimposed image is required (Bernstein 1967; Goodall 1962). The hypothesis also predicts that the constructional capacity of apes will break down at the point at which an operation must be perceptually guided to a location that can only be defined in terms of its future perceptual features or relations: to the part of a stone flake that will become the point of an arrowhead, to the part of a stick that will be the left-hand side of a notch, or to the part of the cabinet that will be obscured when the hinges are in place. The experiments with apes indicate that their ability to impose form upon constructions is extremely rudimentary or nonexistent. Apes, for example, can hold a paintbrush and grasp the relationship between placing the brush in paint and subsequent color upon the paper, but in none of the ape "art" thus produced is there evidence of systematic pattern (Glaser 1971; Morris 1962). Similarly, if a "blank" outline of a chimpanzee face is made out of a photograph and the paper cutouts of eyes, nose, and mouth are given to a chimpanzee, even conceptually sophisticated animals do not arrange the pieces into a face pattern, even though children do so with alacrity (Premack 1975). This inability to externalize form is all the more striking because it coexists in animals with proven ability to recognize pictures in books, to utilize names for concepts, and to perform complex sequences of manipulative tasks under the control of conditional relationships among conceptual exemplars.

However, apes are in fact transitional between monkeys and humans in constructional ability if one accepts the possibility that the control of motor activity by internal imagery begins not with objects but with the body itself. Both gorillas and chimpanzees have been taught language analogs that utilize specific configurations of the hand to convey specific concepts (Patterson 1978; Gardner and Gardner 1971; Fouts 1975). In American Sign Language, the signs for concepts are encoded by a combination of (1) hand shape or configuration, (2) the location at which the sign is articulated, (3) the movements that interrelate the hand configuration and location, and (4) the orientation of the hands. For example, the ASL signs for *name* and *egg* have the same hand shape, the same orientation, and the same location, but they differ in the movement of the hands. In contrast, the signs for *bird* and *print* share hand shape, orientation, and movement but differ in the place of articulation. Still other signs require different hand configurations. The sign for *oil* is made with thumb and index finger touching, but *look* has the index finger extended from a compact hand. Sign language in apes is usually thought of as a process of communication, but it is just as much a process of construction. Specific body parts are being related to other body parts by particular patterns of movement and orientation, analogous to the perceptually controlled movements needed in construction, and the different hand configurations provide an exact parallel to the phenomenon of a motor act being driven by an internal image. When an ape produces a string of distinct hand shapes as part of a spontaneous utterance in sign language, the conformation of the hands must be generated from internally stored information. Moreover, in the teaching of sign language to apes, the most efficacious process was not the rewarding of spontaneous hand configurations that approximated signs in ASL, but a molding process in which the experimenter took the animal's hand into his own and physically formed it into the desired configuration (Fouts 1972). This suggests an iconic process, either visual or kinesthetic, in which the shape is being subsequently used as the goal of the hand movements.[7]

The ability of apes to use signs, if viewed as a constructional process rather than as a specifically communicative process, is compatible with the other differences in primate constructional action already noted. Monkeys can use internally generated images to locate exemplars in the external world, as with Davis's macaques locating circles of an exact size, and they can manipulate objects using current perceptual information. Apes and probably *Cebus* monkeys can manipulate objects and interrelate them in configurations relative to their own body and also to other objects to make *constructions*. In addition, apes can use internally generated images to control the configurations of body parts. Hominids can do all these things but also can use internally generated images to relate one object to another in the making of constructions. The phylogenetic continuity of primate cognition gives rise to a discontinuity once the hominid grade emerges. As iconic control of the motor system moves distalward, from the eye to the hand to external objects, the possibility arises of image-driven productions that remain after their makers have ceased to operate on them. Although these productions are generally referred to as culture — specifically material culture — the discontinuity is primarily one of result rather than of process. The constructional ability of hominids, although discontinuous with apes in one respect, can also be seen as the next logical step in the distalward migration of iconic motor control, a progression with an ancient history within the order. Stated in another way, *the emergence of a hominid level of function requires the incorporation of programs that control the extensions of the body into the instrumental programs that already control the body itself.*

If this argument is accepted, then contemporary nonhuman primates are far more similar to modern man than has generally been recognized. The profound behavioral discontinuity, revealed in human constructional activity, results from the evolutionary elaboration of object-using programs, but the underlying instrumental control and its relationship to the affective modality remains unchanged. It follows that if humans do with objects what apes do with their hands, then societies of humans and apes are biologically homologous at the level of psychological mediation however dissimilar their behavioral features. Consequently, contemporary nonhuman primates can be used as models of hominid society before the elaboration and integration of object-using programs. By this simple strategy, which follows logically from the postulated homology of the processing mechanisms of the primate brain and the postulated change in the evolution of instrumentality, it can be shown that the social organization of manual activity of nonhuman primates — specifically social grooming — has many of the same properties social science theorists have attributed to the emergence of a cultural level of organization in man. Since these same theorists are already in agreement that nonhuman primates are lacking a cultural level of organization, dualistic social science is faced with a choice between two equally unpalatable conclusions: that many properties of human society can emerge in the absence of culture, and hence are innate or self-taught, or that nonhuman primates have a cultural level of organization in many ways comparable to man. Either way, monkeys are hard to ignore.

SUMMARY

Experimental and comparative studies of primate cognition argue for a progressive evolution of conceptual thought and the constructional use of objects, especially when humans are compared to apes and apes to most monkeys. However, studies of human constructional activity also indicate a phylogenetic discontinuity in the human ability to direct motor acts in terms of an image of a final product that is being produced by the action. Sign language, in which internally generated hand shapes are used instrumentally by apes, provides a transitional case to this human capacity for image-driven object construction: apes can iconically manipulate their bodies, whereas humans can iconically manipulate both their bodies and external constructions. Consequently, many of the behavioral differences between apes and humans can be explained as progressive changes in instrumentality that allow external objects to fill the role now occupied by the body in apes. In the next chapter, it will be argued that one program of dextrous manipulation—social grooming—is largely replaced by object manipulation and construction in the transition to hominids but that the underlying motivational dynamics remain the same.

NOTES TO CHAPTER 4

1. Models of conceptualization have been most explicitly developed by computer simulation techniques (Anderson 1976, Bobrow and Collins 1975, Minsky 1975, Schank and Colby 1976).

2. See also Harlow's work in Riopelle (1967), as well as Hayes et al. (1953), Rohles and Devine (1967), Hicks (1956), and Meyer (1971).

3. Also Pribram (see chap. 2).

4. See chap. 3.

5. Also P. C. Reynolds (personal observation).

6. See reference to Evarts, chap. 2. However, internal imagery is not necessarily programmed by the visual system, though it may presuppose it.

7. Terrace's claim that ape sign language is deficient in the mutual "expansion" of utterances, in which each participant builds on what the other has just said, is in complete accord with the constructional hypothesis (1979). Ape language fails at precisely the point at which apes fail more generally: in the cooperative construction of a joint product.

REFERENCES

ANDERSON, JOHN R. 1976. *Language, memory and thought.* Hillsdale, N.J.: Erlbaum.

ANONYMOUS 1976. *Kites to make and fly.* Hammondsworth, Eng.: Kestrel Books, Penguin.

BARGE, E. M., Jr., and THOMAS, R. K. 1969. Conditional discrimination in the squirrel monkey. *Psycho. Sci.* 15:161–163.

BECK, BENJAMIN B. 1973. Observation learning of tool use by captive guinea baboons *(Papio papio). Amer. J. Physical Anthropology* 30:579–582.

  1974. Baboons, chimpanzees, and tools. *J. Human Evolution* 3:509–516.

  1975. Primate tool behavior. In *Socioecology and psychology of primates,* ed. R. H. Tuttle, pp. 413–447. The Hague: Mouton.

BERNSTEIN, I. S. 1961. The utilization of visual cues in dimension-abstracted oddity by primates. *J. Comp. Physiol. Psychol.* 54:243–247.

  1967. Age and experience in chimpanzee nest building. *Psychol. Rep.* 20:1106.

BERTRAND, M. 1967. Training without reward: traditional training of pig-tailed macaques as coconut harvesters. *Science* 155:484–486.

BIRCH, HERBERT G. 1945. The relation of previous experience to insightful problem-solving. *J. Comp. Psychol.* 38:367–383.

BOBROW, DANIEL G., and COLLINS, ALLAN, eds. 1975. *Representation and understanding: studies in cognitive science.* New York, San Francisco, and London: Academic Press.

BOLWIG, N. 1961. An intelligent tool-using baboon. *South African J. Sci.* 57:147–152.

  1963. Observations on the mental and manipulative abilities of a captive baboon *(Papio doguera). Behaviour* 22:24–39.

BORDES, FRANÇOIS 1971. Physical evolution and technological evolution in man: a parallelism. *World Archaeology* 3:1–5.

BROWN, W. LYNN; OVERALL, J. E.; and BLODGETT, H. C. 1959. Novelty learning sets in rhesus monkeys. *J. Comp. and Physiol. Psych.* 52:330–332.

BRUNER, J. S.; GOODNOW, J.; and AUSTIN, G. A. 1956. *A study of thinking.* New York: Wiley.

CHAMBERS, JAMES C., Jr., and TAVUCHIS, NICHOLAS 1976. Kids and kin: children's understanding of American kin terms. *J. Child Language* 3:63–80.

CHIANG, M. 1967. Use of tools by wild macaque monkeys in Singapore. *Nature* (London) 214:1258–1259.

COWEY, ALAN, and WEISKRANTZ, LAWRENCE 1975. Demonstration of cross-modal matching in rhesus monkeys, *Macaca mulatta*. *Neuropsychologica* 13:117–120.

1976. Auditory sequence discrimination in *Macaca mulatta:* role of superior temporal cortex. *Neuropsychologica* 14:1–10.

DAVENPORT, RICHARD K., and ROGERS, CHARLES M. 1970. Intermodal equivalence of stimuli in apes. *Science* 168:279–280.

1971. Perception of photographs by apes. *Behavior* 39:318–320.

DAVENPORT, R. K.; ROGERS, C. M.; and RUSSELL, I. S. 1973. Cross modal perception in apes. *Neuropsychologica* 11:21–28.

DAVIS, R. T. 1971. Memory for form by monkeys. In *Proceedings of Third International Congress on Primatology,* ed. H. Kummer, vol. 3, pp. 150–157. Basel: Karger.

DEAG, J. M., and CROOK, J. H. 1971. Social behavior and "agonistic buffering" in the wild barbary macaque *Macaca sylvana* L. *Folia Primatol.* 15:183–200.

DEWSON, J. H. 1977. Preliminary evidence of hemispheric asymmetry of auditory function in monkeys. In *Lateralization in the nervous system,* ed. S. Harnad, R. W. Doty, L. Goldstein, J. Jaynes, and G. Krauthamer, pp. 63–71. New York: Academic Press.

DÖHL, J. 1966. Manipulier Fahigkeit und "einsichtiges" Verhalten eines Schimpansen bei komplizierten Handlungsketten. *K. Tierpsychol.* 23:77–113.

EATON, GRAY 1972. Snowball construction by a feral troop of Japanese macaques *(Macaca fuscata)* living under seminatural conditions. *Primates* 13:411–414.

EKMAN, PAUL, and FRIESEN, WALLACE 1969. Nonverbal leakage and clues to deception. *Psychiatry* 32:88–106.

EKMAN, PAUL; FRIESEN, WALLACE V.; and SCHERER, KLAUS R. 1976. Body movement and voice pitch in deceptive interaction. *Semiotica* 16:23–27.

ERWIN, J. 1974. Laboratory-reared rhesus monkeys can use their tails as tools. *Percept. and Mot. Skills* 39:129–130.

ETTLINGER, G., and JARVIS, M. J. 1976. Cross-modal transfer in the chimpanzee. *Nature* (London) 259:44–46.

FOUTS, R. S. 1972. Use of guidance in teaching sign language to a chimpanzee *(Pan troglodytes). J. Comp. and Physiol. Psychol.* 80:515–522.

1975. Capacities for language in great apes. In *Socioecology and psychology of primates,* ed. R. Tuttle, pp. 371–390. The Hague: Mouton.

FRENCH, GILBERT M. 1965. Associative problems. In *Behavior of nonhuman pri-*

mates: modern research trends, ed. Allan M. Schrier, Harry F. Harlow, and Fred Stollnitz, vol. 1. New York: Academic Press.

GARDNER, B. T., and GARDNER, R. A. 1971. Two-way communication with an infant chimpanzee. In *Behavior of nonhuman primates: modern research trends,* ed. A. M. Schrier and F. Stollnitz, vol 4, pp. 117–184. New York and London: Academic Press.

GLASER, H. S. R. 1971. Differentiation of scribbling in a chimpanzee. In *Proc. Third Internat'l. Cong. Primatology,* ed. H. Kummer, vol. 3, *Behavior,* pp. 142–149. Basel: Karger.

GOODALL, JANE 1962. Nest-building behavior in the free-ranging chimpanzee. *Annals N.Y. Acad. Sciences* 102, art. 2: 455–467.

HAMILTON, W. J., III; BUSKIRK, R. E.; and BUSKIRK, W. H. 1975. Defensive stoning by baboons. *Nature* (London) 256:488–489.

HAYES, KEITH J., and NISSEN, CATHERINE H. 1971. Higher mental functions of a home-raised chimpanzee. In *Behavior of nonhuman primates: modern research trends,* ed. A. M. Schrier and F. Stollnitz, vol. 4, pp. 59–115. New York and London: Academic Press.

HAYES, K. J.; THOMPSON, R.; and HAYES, C. 1953. Concurrent discrimination learning in chimpanzees. *J. Comp. and Physiol. Psychol.* 46:105–107.

HICKS, LESLIE H. 1956. An analysis of number-concept formation in the rhesus monkey. *J. Comp. and Physiological Psychol.* 49:212–218.

HOLLOWAY, RALPH 1969. Culture: a human domain. *Current Anthropology* 10:395–412.

JONES, C., and SABATER PI, J. 1969. Sticks used by chimpanzees in Rio Muni, West Africa. *Nature* (London) 223:100–101.

KAWAI, M. 1975. Precultural behavior of the Japanese monkey. In *Hominisation and verhalten,* ed. G. Kurth and I. Eibl-Eibesfeldt, pp. 32–55. Stuttgart: Gustav Fischer Verlag.

KINTZ, B. L.; FOSTER, M. S.; HART, J. O.; OMALLEY, J. J.; PALMER, E. L.; and SULLIVAN, S. L. 1969. A comparison of learning sets in humans, primates and subprimates. *J. Gen. Psychol.* 80:189–204.

KOHLER, WOLFGANG 1959. *The mentality of apes.* New York: Vintage Books. 1st Eng. ed., 1925.

KUMMER, HANS 1967. Tripartite relations in hamadryas baboons. In *Social communication among primates,* ed. S. Altmann, pp. 63–71. Chicago: U. of Chicago Press.

LADYGINA-KOTS, N. N., and DEMBOVSKII, Y. N. 1969. The psychology of primates. In *A handbook of contemporary Soviet psychology,* ed. M. Cole and I. Maltzman. New York: Basic Books.

LAWICK-GOODALL, J. van; LAWICK, H. van; and PACKER, C. 1973. Tool-use in free-living baboons in the Gombe National Park, Tanzania. *Nature* (London) 241:212–213.

LEHR, E. 1967. Experimentelle untersuchen an Affen und Halbaffen über generalisation von Insekten-und Blütenabbildungen. *Z. Tierpsychol.* 24:208–244.

McGREW, W. C. 1974. Tool use by wild chimpanzees in feeding upon driver ants. *J. Human Evolution* 3:501–508.

MASSAR, R. S., and DAVIS, R. T. 1959. The formation of a temporal-sequence learning set by monkeys. *J. Comp. and Physiol. Psychol.* 52:225–227.

MENZEL, E. W., Jr. 1971. Communication about the environment in a group of young chimpanzees. *Folia Primatol.* 52:220–232.

1972. Spontaneous invention of ladders in a group of young chimpanzees. *Folia Primatol.* 17:87–106.

1973. Chimpanzee spatial memory organization. *Science* 182:943–945.

MENZEL, E. W., Jr.; DAVENPORT, R. K.; and ROGERS, C. M. 1970. The development of tool using in wild-born and restriction-reared chimpanzees. *Folia Primatol.* 12:273–283.

MERZ, ELLEN 1978. Male-male interactions with dead infants in *Macaca sylvanus*. *Primates* 19:749–754.

MEYER, DONALD R. 1971. The habits and concepts of monkeys. In *Cognitive processes of nonhuman primates,* ed. L. E. Jarrard, pp. 83–102. New York: Academic Press.

MILES, RAYMOND C. 1965. Discrimination-learning sets. In *Behavior of nonhuman primates: modern research trends,* ed. A. Schrier, H. Harlow, and F. Stollnitz, vol. 1, pp. 51–95. New York: Academic Press.

MINSKY, MARVIN 1975. A framework for representing knowledge. In *The psychology of computer vision,* ed. Patrick Henry Winston, pp. 211–277. New York: McGraw-Hill.

MORRIS, DESMOND 1962. *The biology of art.* New York: Knopf.

NEWMAN, J. D., and SYMMES, D. 1974. Arousal effects on unit responsiveness to vocalizations in squirrel monkey auditory cortex. *Brain Research* 78:125–138.

NISHIDA, TOSHISADA 1973. The ant-gathering behaviour by the use of tools among wild chimpanzees of the Mahali Mountains. *J. Human Evolution* 2:357–370.

NISSEN, H. W. 1951. Analysis of complex conditional reaction in chimpanzee. *J. Comp. and Physiol. Psychol.* 44:9–16.

NISSEN, H. W.; BLUM, J. S.; and BLUM, R. A. 1948. Analysis of matching behavior in chimpanzee. *J. Comp. and Physiol. Psychol.* 4:62–74.

1949. Conditional matching behavior in chimpanzee; implications for the comparative study of intelligence. *J. Comp. and Physiol. Psychol.* 42:339–356.

PARKER, C. E. 1974. The antecedents of man the manipulator. *J. Human Evolution* 3:493–500.

PATTERSON, F. 1978. The gestures of a gorilla: sign language acquisition in another pongid species. *Brain and Language* 5:72–94.

PLOOG, DETLEV 1967. The behavior of squirrel monkeys *(Samiri sciureus)* as revealed by sociometry, bioacoustics, and brain stimulation. In *Social communication among primates,* ed. Stuart A. Altmann, pp. 149–184. Chicago: U. of Chicago Press.

PREMACK, ANN JAMES, and PREMACK, DAVID 1972. Teaching language to an ape. *Scientific American,* Oct., pp. 92–99.

PREMACK, DAVID 1975. Putting a face together. *Science* 188:228–236.

1976. *Intelligence in ape and man.* Hillsdale, N.J.: Erlbaum.

RENSCH, B., and DÖHL, J. 1967. Spontanes Öffnen verschiedener Kistenverschlüsse durch einen Schimpansen. *Z. Tierpsychol.* 24:476–489.

RIOPELLE, A.J. ed. 1967. *Animal problem solving.* Baltimore: Penguin Books.

RIOPELLE, A. J., and COPELAN, ELTON L. 1954. Discrimination reversal to a sign. *J. Experimental Psych.* 48:143–145.

ROBINSON, BRYAN W. 1967. Vocalization evoked from forebrain in *Macaca mulatta. Physiology and Behavior* 2:345–354.

ROGERS, CHARLES M., and DAVENPORT, RICHARD K. 1975. Capacities of nonhuman primates for perceptual integration across sensory modalities. In *Socioecology and psychology of primates,* ed. R. Tuttle, pp. 343–352. The Hague: Mouton.

ROHLES, F. H., JR., and DEVINE, J. V. 1967. Further studies of the middleness concept with the chimpanzee. *Animal Behaviour* 15:107–112.

RUMBAUGH, DUANE M. 1970. Learning skills of anthropoids. In *Primate behavior: developments in field and laboratory research,* ed. Leonard A. Rosenblum, vol. 1, pp. 1–70. New York: Academic Press.

1975. The learning and symbolizing capacities of apes and monkeys. In *Socioecology and psychology of primates,* ed. R. Tuttle, pp. 353–365. The Hague: Mouton.

RUMBAUGH, DUANE M., and GILL, TIMOTHY V. 1973. The learning skills of great apes. *J. Human Evolution* 2:171–179.

SABATER PI, JORGE 1974. An elementary industry of the chimpanzees in the Okorobiko Mountains, Rio Muni (Republic of Equatorial Guinea), West Africa. *Primates* 15, no. 4:351–364.

SAVAGE-RUMBAUGH, E. SUE; RUMBAUGH, D. M.; and BOYSEN, SALLY forthcoming. Linguistically-mediated tool use and exchange. *Behavioral and Brain Sciences.*

SCHANK, ROGER C., and COLBY, KENNETH MARK, eds. 1973. *Computer models of thought and language.* San Francisco: Freeman.

SCHILLER, P. H. 1952. Innate constituents of complex responses in primates. *Psychological Review* 59:177–191.

SCHILLER, P. H. 1957. Innate motor action as a basis of learning. In *Instinctive behavior: the development of a modern concept,* ed. C. H. Schiller, pp. 264–287. London: Methuen.

SINNOTT, J. M.; STEBBINS, W. C.; and MOODY, D. B. 1973. Regulation of voice amplitude by the monkey *(Macaca). J. Acoust. Soc. Amer.* 53:378.

STRONG, PASCHAL N. 1959. Memory for object discriminations in the rhesus monkeys. *J. Comp. and Physiol. Psychol.* 52:333–335.

STRUHSAKER, T. T., and HUNKELER, P. 1971. Evidence of tool-using by chimpanzees in the Ivory Coast. *Folia Primatol.* 15:212–219.

SUTTON, D.; LARSON, C.; and LINDEMAN, R. C. 1974. Neocortical and limbic lesion effects on primate phonation. *Brain Research* 71:61–75.

TALMADGE-RIGGS, G.; WINTER, P.; and MAYER, W. 1972. Effects of deafening on the vocal behavior of the squirrel monkey *(Saimiri sciureus). Folia Primatol.* 17:404–420.

TELEKI, G. 1974. Chimpanzee subsistence technology: materials and skills. *J. Human Evolution* 3:575–594.

TEMERLIN, MAURICE K. 1975. *Lucy: growing up human.* Palo Alto, Calif.: Science and Behavior Books.

TERRACE, H. S. et al. 1979. Can an ape create a sentence? *Science* 206:891–902.

TYLER, STEPHEN A., ed. 1969. *Cognitive anthropology.* New York: Holt, Rinehart and Winston.

VEVERS, G. M., and WEINER, J. S. 1963. Use of a tool by a captive capuchin monkey *(Cebus apella).* In *The primates; Symposia Zoological Soc. London,* ed. J. Napier and N. A. Barnicot, no. 10:115–117.

WARREN, J. M. 1965. Primate learning in comparative perspective. In *Behavior of nonhuman primates: modern research trends,* ed. Allan M. Schrier, Harry F. Harlow, and Fred Stollnitz, vol. 1, pp. 249–281. New York: Academic Press.

WOLLBERG, ZUI, and NEWMAN, JOHN D. 1972. Auditory cortex of squirrel monkey: response patterns of single cells to species-specific vocalizations. *Science* 175:212–214.

WRIGHT, R. V. 1972. Imitative learning of a flaked stone technology: the case of an orangutan. *Mankind* 8:296–306.

YOUNG, M. L., and HARLOW, H. F. 1943a. Solution by rhesus monkeys of a problem involving the Weigl-principle using the oddity method. *J. Comp. Psychol.* 35:205–217.

1943b. Generalization by rhesus monkeys of a problem involving the Weigl-principle using the oddity method. *J. Comp. Psychol.* 36:201–216.

5

THE WEALTH OF MONKEYS

INTRODUCTION

In the previous chapter it was argued that human evolution has been characterized by the elaboration of instrumental action—the progressive sophistication of the control programs that operate peripheral effectors, the use of
internal images to drive constructional processes, and the incorporation of
material objects into instrumental action. This perspective views the
emergence of instrumentality as phylogenetically older than the ability to
use material objects in an instrumental way, and it also cautions against
equating tool using with physical tools. Instrumentally controlled innate
behaviors are the first tools, and these behaviors come to be supplemented
by advanced programs for the manipulation and construction of material
objects in the course of human evolution. However, in chapter 3 it was
argued that primates generally are characterized by functional interaction
between affectivity and instrumentality, and ethological studies demonstrate
the affective continuity of man and ape. This not only suggests that humans
use objects as apes use innate behaviors, but it follows that there should be a
basic comparability between primitive object-manipulation programs and
advanced object-manipulation programs *in spite of the dissimilarity of behavioral morphology:* the dissimilar instrumental behavior of humans and apes
should be comparable in the underlying affective control because affectivity
is in many ways comparable in humans and apes. Moreover, the social
dynamics of human and ape object manipulation ought to be similar, because affectivity is a system specialized for the assessment and programming
of social action. Given this theoretical perspective, evolutionary strategy
need only postulate the incorporation of object-manipulation programs of
a human level of complexity into the object-manipulation of apes. That is,

wherever the dextrous manipulation of matter occurs in ape social relations, we can conceptualize the hominid transition as its supplementation by advanced programs for the manipulation of material objects, such as the image-driven constructional action already noted.

If we make this mental substitution, then many of the behaviors of nonhuman primates that have formerly been regarded as inconsequential assume a new theoretical weight. Grooming behavior, for example, in some species the most frequent kind of social action, is not a trivial antecedent to louse picking in man but an antecedent to *social exchanges involving objects*. That is, the grooming *behavior* of monkeys participates as a token in networks of exchange relationships, and human evolution has been characterized by the transformation of *grooming networks* into *object-exchange networks* by the substitution of advanced object-manipulation programs for the simpler instrumental actions of grooming itself. Wealth and economy, in other words, no more emerge with the use of material objects than does instrumentality. Just as a tool is the knowledge and skill that allows a material object to be used instrumentally in goal-directed action, so wealth is a token that can function in systems of social exchange. Even in the case of the material object exchange that characterizes human social relationships, the material token is firmly embedded in purely behavioral programs that implement the actual exchange, either by verbal denotation or actual transfer, and the material object has no exchange value beyond its capacity to engage these behavioral programs. To equate wealth with material objects is to participate in a common reification, discussed by Freud in the psychoanalysis of money, and this equation is scientifically inadequate because it wrenches the material token from its functional context and diverts attention from the exchange process itself. Viewed from the perspective of exchange *processes* while accepting the dissimilarity of exchange *tokens,* the grooming networks of nonhuman primates and the object exchange networks of humans do in fact have a great many properties in common, as an evolutionary transformation model suggests.

THE TWO AXES OF GROOMING

QUIET VERSUS CONFLICT

It has often been noted that grooming is a frequent displacement activity, occurring as seemingly irrelevant behavior in conflict situations or marking behavioral transitions. As Iersel and Bol found in their study of displacement preening in terns (*Sterna* spp.), the preening occurred when the motivations to incubate and to flee were equally balanced: when both were low, both intermediate, or both high, with the greatest amount of grooming given when both were approximately intermediate in strength (Iersel and Bol 1958). Andrew found that preening in chaffinches was associated with behavioral transitions (1956), and Kendon and Ferber found that self-grooming was associated with transitions to new activities in a group of adolescent humans (1973). Similarly, self-grooming is a prominent aspect of behavior in psychiatric interviews, and people self-groom more when social distance is reduced (Scheflen 1973; Kleck 1970). It is also known that self-grooming is systematically related to other forms of self-maintenance activities, such as feeding and sleep (Moruzzi 1969; Delius 1970, Dell et al. 1961; Wisenfeld et al. 1977). Electrostimulation of the reticular formation in the brainstem has long been known to produce systematic behavioral transitions as different neural systems become engaged with increased neural activity, and self-maintenance behaviors are integrated with the endogenous cycles of sleep and wakefulness. Similarly, Delius found that the behavior patterns occurring during displacement activity in gulls also can be elicited in conjunction with self-maintenance activity by electrostimulation of the brain. Also, in cats, sites in the brainstem that elicit self-grooming overlap with those for feeding (Berntson and Hughes 1974). Studies of primate social behavior also indicate a relationship between grooming and conflict situations as well as between grooming and relaxed activity. However, in primates the situation is complicated by the widespread development of social grooming. If self-grooming is phylogenetically prior to social grooming, then the latter can have two distinct evolutionary derivations. It can be either an other-directed form of relaxed hygienic grooming or an other-directed form of displacement grooming. Accordingly, grooming behavior should vary systematically with states of relaxed familiarity on one hand and with conflict states on the other. The social aspect of grooming is also compatible with the concept of social attachments as stabilized approach-avoidance gradients, since this conflict would facilitate the conditions for both displacement and hygienic grooming, depending upon the state of the relationship.

It is well known that grooming behavior in primates is preferentially associated with periods of quiet relaxation such as the hot parts of the day or the early morning, as has been reported for many species. *Lemur catta* and *Propithecus verreauxi* sun themselves upon waking in the morning, and both species engage in self-grooming during this time, using their specially modified incisor teeth ("tooth comb") and specialized second toes on the hind feet ("toilet claws") (Jolly 1966, pp. 36–37, 59, 81). Both species also have a siesta period in hot weather, and both social grooming and self-grooming occur during this time as well as during sunning and feeding periods. *Papio hamadryas* also begin the day with a morning rest period, and adult grooming, copulation, and chasing occur during this time, as does the play of immature animals (Kummer 1968a). Like the diurnal lemurs, there is also a midday rest period, in which grooming and play occur, but neither chasing nor copulation appear at this time. In South African baboons, *Papio ursinus*, there are pronounced grooming peaks very early in the morning or before retiring at night (Hall 1962, fig. 3). Siamangs and gibbons do not take a siesta, but they begin activities early in the morning and become more restful as the day progresses (Chivers 1972, pp. 120–121, 126). Both species, which are pair-bonded, spend time in long social grooming bouts, about fifteen minutes a day in gibbons and over an hour in siamangs. Grooming, although variable among groups, shows a peak in the midday. Among rhesus monkeys in India, grooming reached a peak in midmorning, after feeding, and fell off toward midday as the animals became more restful (Lindburg 1973, pp. 127–128; Drickamer 1976). Grooming also preceded the afternoon activity period. In a study of activity rhythms in colony-housed pigtail macaques, Bernstein found that peak frequency scores for resting, object-manipulation and self-manipulation were obtained in the early morning, whereas social grooming peaked in the late morning and self-directed behavior fell to its lowest level at that time (1972). The relationship of grooming to the activity cycle is well known, but the social aspects of grooming have been more refractory to analysis since there is so much variation among primate species.

Nearly all species of primates manipulate their own fur at one time or another, but even here there are apparent exceptions. The monkey *Callicebus torquatus* has not been observed to self-groom in the field, even though it socially grooms and has been seen to scratch (Kinzey 1977, p. 176). In some species of South American monkeys, *Saimiri* and *Alouatta palliata,* for example, social grooming is not a significant part of the behavioral repertoire. Among species of primates that do socially groom, there are great differences in the frequency, directionality, and contexts of grooming that do not correlate well with either ecological zone or taxonomic grade. Because of this variation, there is no generally accepted terminology for grooming among primatologists. In the best review of the subject, Simonds distinguishes mother-infant grooming, copulatory grooming, displacement grooming, and hygienic grooming, among others (1974). In an earlier review, Sparks distinguished categories of social grooming on the basis of the mutuality of the activity: whether the behavior was sequentially reciprocated, mutually performed, or unilateral (1967). These various categorizations are useful for organizing observational data, but they present terminological difficulties. Since *mutual* and *reciprocal* are to some extent synonomous in ordinary language, Sparks's terminology can be confusing. For the present discussion, it is the directionality of grooming — its net flow from one individual to another — that is of primary concern, not the simultaneity of the behavior within dyadic interactions. For this, we will refer to *asymmetrical* or *symmetrical* grooming flow. Simonds's categorization of grooming types is largely based on the delimitation of the behavioral context in which the grooming occurs, but his definition of displacement grooming employs mixed criteria. Displacement grooming is characterized by "rapid, jerky movements" and aimed at "reducing the groomer's tensions." Here morphological and motivational definitions are combined in a single label. Behavior conforming to this definition has been observed in macaques and baboons, but the ethological approach would regard displacement grooming as far more extensive than the irregular, jerky grooming of manifestly tense situations. The disturbances of the tempo and feedback control of grooming that occur in tense situations can be observed when any ongoing voluntary motor act is subjected to emotional arousal. For example, the same kinds of disruption can be seen in the shoveling activity of children in a sandpit. There is no reason to regard this disrupted grooming of macaques and baboons as coextensive with displacement activity. Displacement grooming is not a type of grooming but a characterization of grooming behavior that occurs as a result of conflicted motivational systems. Given the concept of attachment relationships as stabilized approach-avoidance gradients, a great deal of morphologically normal grooming can be accommodated to the ethological theory of displacement activity.

As Simonds recognizes, the grooming of newborns by mothers, particularly oral contact, should probably be distinguished as a subcategory of maternal grooming elicited by maternal hormone factors, birth fluids, and the stimulus features of the newborn infant. This mother-infant grooming is extremely widespread, occurring in species that do not otherwise socially groom, and it can differ from other social grooming in target area and behavioral morphology. Orangutans rarely groom socially in the wild, but mothers will lick newborn infants, particularly the genitalia (MacKinnon 1974, p. 51; Rijksen 1978, pp. 209, 241). Similarly, Bertrand has pointed out that female stumptail macaques *(Macaca speciosa)* give frequent licking of the anogenital area of their infants during the first week of life, but that subsequent licking is directed to other parts of the body (1969). Horwich observed in langurs that maternal grooming had more oral behavior than the grooming of noninfants (1974). Moreover, most social grooming in this species is not licking behavior but manipulation of the fur with the hands. In *Galago senegalensis,* a sudden onset of intensive grooming, both social and self-directed, can precede parturition, and the first few hours after birth are usually occupied with intensive grooming of the neonate by the mother (Doyle et al. 1967, 1969). In *Lemur,* too, grooming is associated with the birth of the baby (Vick and Conley 1976, p. 138; Klopfer and Klopfer 1970). In some species of mammals, such as rats, the licking of the anogenital area by the mother is necessary for the survival of the infant, and such oral mother-infant grooming may be part of the mammalian maternal complex, accessible through a specific pathway. Even so, the maternal dyad is not incompatible with the idea of a stabilized approach-avoidance conflict containing its own genetically encoded form of other-directed displacement grooming.

This motivational typology of grooming is based on distinctions among the hypothetical mechanisms that control the overt behavior, but it predicts observable systematic relationships between grooming and other features of behavior, such as the sleep–wake cycle, sex, age, familiarity, status, and emotional tension. Social grooming derived from hygienic maintenance behavior should occur in temporal proximity to sleep and be directed to familiar or intimate individuals. The studies of isolation rearing already discussed indicate that many social behaviors, such as sexuality and aggression, which are normally directed toward other individuals, can become self-directed in the absence of social stimulation. The same processes of social interaction and social familiarity can perhaps facilitate the converse effect and extend self-directed behaviors, such as grooming, toward other individuals. Whatever the actual mechanism, there is no doubt that other-directed grooming has developed in a number of species of birds and mammals. Since maintenance activities take place in quiet and familiar circumstances, any extension of hygienic grooming would presuppose familiar individuals who would not disrupt the relaxed motivational state. Social grooming derived from the maternal care complex should be recognizable by its cooccurrence with other forms of maternal care giving, such as carrying, protecting, and feeding the young; and the social target of the behavior should be morphologically more infantile than the agent. Moreover, *maternal* or *care-giver* grooming can be recognized by a characteristic asymmetry in directionality. It is not dependent upon reciprocity but flows unilaterally from the older individual to the younger.

CARE-GIVERS GROOM MORE

If some social grooming is derived from the maternal care complex, then this grooming should relate systematically to the age-sex category which serves as the primary care giver to infants. In an early publication, Rosenblum, Kaufman, and Stynes noted a tendency of primate females to groom more than males (1966).[1] This tendency is indeed a widespread one. It has been reported from *Alouatta seniculus, Cebus capucinus, Cebus nigrivittatus, Colobus badius, Colobus guereza, Cercopithecus campbelli, Cercopithecus mitis, Cercopithecus aethiops. Miopithecus talapoin, Cercocebus atys, Cercocebus albigena, Erythrocebus patas, Presbytis entellus,* for all species of *Macaca* except *radiata* in some circumstances, and for all *Papio,* although in *hamadryas* and *anubis* male grooming may approach parity with females (Oppenheimer and Oppenheimer 1973; and below). In all of these species, except *Macaca sylvanus,* females are the primary care givers, and the higher level of female grooming is compatible with this role. If social grooming has a maternal component, then males ought to groom more than females in species where the male is the primary care giver. This, in fact, is generally the case for the paternal-care species that have been studied. Greater *male* grooming has been reported for *Callithrix, Aotus, Callicebus torquatus,* and *Symphalangus,* all of which have males that carry the offspring (Rothe 1971). However, the

176

male grooming in such species is not necessarily directed to the offspring, as in normal care-giver grooming, but may be directed to the partner. Females are also more "groomy" to many age-sex categories when they are primary care givers. In a *Symphalangus* troop observed by Chivers, grooming was primarily between the male and female, with the male grooming twice as much, but the male also groomed the subadult in the troop more than the female did (1974).

Conversely, in species where the females are the primary care givers, males ought to be more peripheralized and less maternal. Accordingly, male grooming of others, especially infants, ought to be rare in such species. Recent field research has shown that there is a continuum of paternal care, ranging from socially aloof males, such as *Presbytis johnii*, which virtually ignore infants and orient primarily to outgroup males, to very paternal males that perform all child rearing except nursing (Mitchell and Brandt 1972). Grooming of infants by males does roughly correspond to the place on the paternal continuum. In some species, like *Colobus guereza* or *Alouatta seniculus*, the adult males almost never groom anyone, even though social grooming is common (Leskes and Acheson 1971; Neville 1972). In *Presbytis entellus*, males do not groom infants, even though some male grooming occurs (Jay 1962). Rhesus monkey males, which are among the least paternal members of *Macaca*, only groom infants infrequently; and in a long-term study, Hinde and his colleagues saw only six examples (1964, p. 644). In the Takasakiyama troop of Japanese macaques *(Macaca fuscata)*, in contrast, adult males will take an interest in infants and carry and groom them (Itani 1959). Also, in Barbary macaques *(Macaca sylvana)*, where paternal behavior is elaborated, males both carry infants and groom them (Simonds 1965, table 6.5). In *Macaca radiata*, adult males have been observed to groom all age-sex categories, but they groom infants and juveniles infrequently. In *Papio hamadryas* and *Theropithecus gelada*, where male care giving toward immature females is a normal stage in the formation of the unimale group, male grooming of the partner is frequent (Kummer 1968a, 1968b; Dunbar and Dunbar 1975). Also, in gibbons, where the female carries the infant but the male is pair-bonded, mutual grooming is high. In East African baboons, *Papio anubis*, adult males are often involved in vigilance behavior toward children; and adoptions, holding, and grooming of infants occur (Devore 1963; Ransom and Ransom 1971). These observations do suggest a relationship between the extent of male paternal involvement with immatures and the tendency to groom them. The species with the most frequent reports of adult males grooming immatures are also generally those species that exhibit other forms of paternal contact behavior. This suggests that in an age-sex category with an orientation toward infants, the threshold for grooming should generally be lower and more accessible in a variety of contexts. One characteristic of care-giver grooming is that it tends to be asymmetrical, since it is given to individuals who are socially and physically dependent. This, too, is frequently observable, as shown by a brief review.

In *Colobus badius,* infants, which groom little, are themselves groomed by their mothers; and in a study of *Colobus guereza (abyssinicus),* infants received 17 percent of the grooming, though they performed less than 1 percent (Leskes and Acheson 1971, p. 25; Struhsaker 1975, p. 50). *Theropithecus gelada* is instructive in this respect because all grooming is reciprocated *except* grooming of young by their mothers (Dunbar and Dunbar 1975). Moreover, care-giver grooming can vary independently of other types, and it can be infrequent in grooming species and common in non-grooming species. In *Cercopithecus aethiops sabaeus,* infants receive more grooming than they give, but it has been noted by several authors that there is little maternal grooming in this species compared to that among macaques (Dunbar 1974; Gartlan 1969; Lancaster 1971; Struhsaker 1971). In *Miopithecus,* in contrast, infants receive the largest portion of female grooming (Gautier-Hion 1971, pp. 323–326). Just the opposite situation is apparently found in *Cercocebus albigena,* where only 5 of 113 grooming bouts observed were between mother and infant, even though social grooming is the second most frequent social behavior after "sitting in contact" (Chalmers 1968; Chalmers and Rowell 1971). In *Presbytis johnii,* grooming is infrequent when compared to *Presbytis entellus,* and maternal grooming is even less frequent (Poirier 1968, p. 56). Only 7 percent of grooming was between a mother and an infant. In *Presbytis entellus,* the mother licks and grooms the infant from birth, and other females also groom and handle the baby (Jay 1965, p. 221). Grooming of neonates has also been reported for *P. obscurus* in captivity, and *P. cristatus* in Malaysia has been described as grooming frequently, with all age-sex classes participating except infants, who receive grooming but do not give it (Horwich 1974; Bernstein 1968). In *Ateles geoffroyi,* social grooming occupied only 5 percent of observed behavior in a captive group, and Carpenter reported only grooming of infants by their mothers (1935).[2] *Gorilla gorilla beringei* is also similar in this respect, with little social grooming, but, of that, over half is directed from mother to offspring (Schaller 1963, p. 245). In chimpanzees, *Pan troglodytes,* grooming has a relatively restricted distribution, occurring primarily between mother and child or among males (Lawick-Goodall 1967). In *Symphalangus,* the mother will groom the infant as they prepare for sleep, but the child does not reciprocate (Chivers 1974).

Other kinds of behavioral evidence also suggest a relationship between the care-giving complex and the frequency of social grooming. In *Lemur catta,* which has frequent two-way grooming in both sexes, the birth of infants increases grooming behavior (Jolly 1966, pp. 95, 117). Other individuals frequently attempt to groom neonates and, during the birth season, adult female grooming of a female's own or other infants is the predominant form. Similarly, in caged *Callithrix jacchus,* grooming between adult females increases to about twice normal levels in the presence of neonates (Rothe 1971). Also, in *Erythrocebus patas,* grooming increases when infants are reunited with their mothers (Preston et al. 1970). *Macaca mulatta* (rhesus) mothers groom their infants soon after birth, and they also groom with the mouth, licking the newborn, particularly the hands and feet (Hinde et al. 1964, pp. 636–637). Hinde and his colleagues found that rate of grooming was constant over the first six months, about 4 percent to 9 percent of total behavior. Hansen, however, reports a statistically significant decrease both in grooming and in the rate of decrease of mother-infant grooming as a function of age during the first three months (1966, p. 117). Both the observation procedures and the group compositions differed in these two studies. These various lines of evidence indicate that the relationship between social grooming and care-giving activity is too systematic for grooming to be simply an extension of hygienic behavior. Moreover, in a study of *Erythrocebus patas* in Cameroon, social grooming declined as the dry season progressed but self-cleaning behaviors remained common (Struhsaker and Gartlan 1970). The target locations of social and hygienic (self) grooming also differ, with animals directing their social grooming to the parts of others that are inaccessible to the self-groomer (Horwich and Wurman 1978; Hutchins and Barash 1976). On the opposite side, in a comparative study of laboratory groups of pigtail *(Macaca nemestrina)* and bonnet *(M. radiata)* macaques, self-grooming and social grooming were to some extent reciprocals of each other (Rosenblum et al. 1966). Significant negative correlations were obtained between the duration of autogrooming and both the initiation and receipt of social grooming. During a brief food shortage in the laboratory, pigtail macaques decreased both autogrooming and social grooming, whereas bonnet macaques reduced their autogrooming but not their social grooming (Rosenblum, Kaufman, and Stynes 1969). Also, when amphetamines are given to rhesus monkeys, both self-grooming and social grooming are reduced (Miller 1976).

Because the evolutionary argument links social grooming and self-maintenance activity, some relationship between the two is to be expected, but the variations among species are so great as to preclude any simple model. Moreover, there is a functional connection between other-directed hygienic grooming and the social grooming linked to maternal care: maternal care always creates familiarity, which in turn would allow the expression of socially directed hygienic grooming. In the care-giver–infant relationship, maternal grooming and hygienic grooming would be so intertwined empirically that it is doubtful the two could be teased apart by observational methods alone. However, detailed studies of the transitions between self-grooming and social grooming might be expected to differentiate partially between maternal and hygienic grooming if this distinction is valid. Functionally, however, care-giving grooming and social hygienic grooming are mutually supportive, since both provide a rewarding and relaxing form of social interaction to individuals who have had a past history of positive interaction. Other-directed hygienic grooming and maternal care are both a product and a cause of what will be termed *relaxed familiarity,* and both differ in their relational properties from displacement grooming.

Social grooming derived from displacement activity would occur in situations that were conflictive rather than relaxed and with individuals who were relative strangers to each other or who had a history of negative social interactions. Moreover, displacement grooming would be expected to flow from the more conflicted to the less conflicted, and, in consequence, it would bear a systematic relationship to age and rank that is exactly the reverse of maternal grooming. Ordinarily, displacement grooming would flow from the lower ranking individual to the higher and from the younger to the older, since these individuals would be expected to be most conflicted in their approach-avoidance. Rank and age are themselves systematically related in primates, with the older usually outranking the younger. Maternal and hygienic grooming will tend to flow down the ranking system, and displacement grooming will tend to flow up. Sparks suggests that elaborated social grooming and strong ranking systems are related to each other, with rank-oriented species grooming more frequently (1967, p. 158). Both relaxed-familiar grooming and displacement grooming should be systematically related to social rank, but with the former the association with high rank is an artifact of the greater age of care givers, whereas displacement grooming is a product of the status disparity itself. In consequence, relaxed-familiar grooming should be relatively unaffected by changes in social rank between the partners, whereas the directionality of displacement grooming should be reversed by any rank reversal. Stated in another way, the social rank correlations with relaxed-familiar grooming should be more closely tied to long-term, life-cycle changes in social status, whereas the rank effects on displacement grooming should respond to short-term dominance interactions. Hence two distinct directions of grooming flow are postulated, based on familiarity and displacement. The field studies of nonhuman primate societies have confirmed rank-related effects on social grooming in the best studied species, namely chimpanzees, rhesus and Japanese monkeys, and savanna baboons. These societies all have two axes of grooming flow as the relaxed familiarity and displacement axes suggest. However, in all of these societies there is also a sexual effect on grooming behavior, with adult male-female grooming increasing when females are sexually receptive. *Sexual grooming,* as Simonds terms it, is theoretically important because it provides a way of negating rank-related asymmetries in grooming flow in species where one sex outranks the other. As such, it can mimic the directionality of care-giver grooming and flow from the higher to the lower.

Among rhesus monkeys *(Macaca mulatta),* the most studied primate except for man, social grooming is common, accounting for 24 percent to 34 percent of adult activity in captive animals (Bernstein and Mason 1963; Southwick 1967). Detailed observations demonstrate the importance of genealogical ties, male rank, and sexual attraction in the grooming of this species. Kaufmann, in a study of the troops originally introduced to a Caribbean island, Cayo Santiago, by C. R. Carpenter in 1942, reports that most grooming involving adult males is with adult females as a partner (1967). This varies from about 50 percent of adult male grooming in the birth season to 88 percent of adult male grooming in the mating season. This is true of high-ranking or middle-ranking males. Low-ranking males mostly groom high-ranking males, but in the mating season all males prefer females. High-ranking and middle-ranking males also groom immatures 12 percent to 21 percent of time (except during the mating season), but these immatures are either close relatives (20 percent), orphans (37 percent) or children of favored female associates (17 percent). Male grooming is strongly related to rank, for in 77 percent of male-male grooming, the subordinate groomed the dominant. Females groom males more than males groom females, and high-ranking females preferentially groom females of their own rank but receive grooming from females of all ranks.

Lindburg in a field study in India reports that grooming among adult males was very rare, with 76 percent of adult male grooming directed to adult females (1971, pp. 95–96). About 50 percent of this grooming occurred in a sexual context. The adult females in these troops also groom the males more than the males groom them. The adult females groom more than any other age-sex class, and most of their grooming is directed to their offspring and to other females. There is some tendency for females to groom females close in rank to themselves. In two troops studied in detail, the highest ranking females groomed the most, but there was no relationship to rank in the amount of grooming received. In fact, 59 percent of total adult female grooming was directed down the hierarchy (Lindburg 1973, pp. 139–140). In one group, the amount of grooming given and received by adult males was a function of their rank, but in the second troop this relationship was confounded by high scores for the third-ranking male. In both groups, the alpha males groomed higher-ranking females more than other females. Sade reports that for Troop F on Cayo Santiago, grooming centricity, as measured by the number of individuals grooming a groomee and the relative centricity of the groomers, correlates with rank among females but not among males (1972). However, the alpha male has the highest grooming centricity among the males, and he is integrated into female grooming cliques. Miller and his colleagues report that, for Troop E on Cayo Santiago, males whose genealogies are known are groomed more by their mothers than by any other class of relatives, with siblings next in frequency (1973). Lindburg reports that 42 percent and 48 percent of the grooming in his two troops was among presumed relatives. In the wild troops described by Lindburg, infants rarely groom but receive a great deal, and yearlings groom but still receive more than they give (1973, p. 136). Donald Sade has applied sociometric methods to the analysis of grooming networks on Cayo Santiago (1972). Taking the relationship "Monkey *a* grooms monkey *b*" as a primary datum and computing the proportion of *a*'s total grooming that *b* receives, Sade mathematically extracted the grooming cliques from analysis of these pairwise relationships. The dominant male participated in grooming cliques with the first and second ranking females of Troop F. Moreover, the four highest ranking females form the core of the grooming network. In a study of males with known genealogies in Troop E on Cayo Santiago, it was found that about half of their social interactions were with relatives (Miller et al. 1973). Of these, related animals performed 66 percent of the grooming. The males groomed mothers most often, followed next by their brothers. Males groomed brothers more than sisters. Grooming and positive interactions are far more frequently directed to relatives than would be expected on the basis of their proportion in the population, whereas aggression is about the rate that would be expected.

Loy and Loy established a social colony of one-year-old and two-year-old rhesus monkeys removed from a free-ranging troop on Cayo Santiago (1974). About two hundred hours of observation of the thirty-three animals yielded 1,753 grooming episodes, of which 1,430 were social. Matrilineally related monkeys, comprising 6.6 percent of total pair combinations, accounted for 34.3 percent of the grooming. Middle-ranking and high-ranking monkeys tended to receive more grooming than they gave. The two-year-old animals were more involved in grooming than the yearlings, and females groomed greater numbers of other monkeys than males. During a period on Cayo Santiago when the shipment of provisions did not arrive, Loy observed that the monkeys increased their foraging time but decreased their social interaction time (1970). Rate of grooming decreased by about half. However, grooming with nonrelatives was most affected, dropping from 44.7 percent of total grooming time to 26.1 percent. The proportion of grooming to relatives actually rose from 55.3 percent to 73.9 percent during the famine.

Michael and Herbert observed grooming among male and female rhesus monkeys that had been artificially paired in test cages. The males groomed the females maximally at the midpoint of the female menstrual cycle, while the female grooming of the male was maximal at the end points of the cycle (1963; Michael et al. 1966). Ovariectomy of females reduces male grooming and abolishes the cyclicity. Injection of estrogen into ovariectomized females has the effect of inducing males to groom them at levels appropriate to midcycle. The effect of ovariectomy was primarily a reduction in the number of accepted invitations (solicits) to groom, and the administration of estrogen increased the number of grooming invitations (solicits) responded to by the males. Vessey removed individual rhesus monkeys from their troops and then released them at variable intervals (1971). Those animals that rejoined their troops attacked the animals that ranked below them and groomed those that were higher.

The Japanese macaque, *Macaca fuscata,* shows many social similarities to rhesus. In a study of a troop with known genealogies, where grooming was scored over several months, 57.6 percent of total grooming occurred among relatives, of which 24 percent was by mothers to daughters, 13 percent by daughters to mothers, 14 percent by mothers to sons, and only 1 percent by sons to mothers (Oki and Maeda 1973, table 1, pp. 152–153; Alexander and Bowers 1969). Only 4.3 percent was between siblings. As the young mature, they groom their mothers more. This reaches a high level in female offspring and remains constant, while grooming of mothers by sons peaks at three years of age and then declines dramatically. Of the grooming among nonfamily members, 73.4 percent were directed from the lower-ranking to the higher-ranking matriline, and females primarily groomed those closest to them in rank. Grooming among adult males was rare, only 9 incidents out of 1,850 observed, but of these, 7 were from subordinate to dominant. The males in the central hierarchy were groomed frequently but did not groom others very often. They were most often groomed by a small number of females, usually of high rank, and there were favored partners among males and females. The highest ranking males did not groom each other. In adult male–adult female grooming, 67.8 percent was from female to male. Sexual grooming occurs in Japanese macaques, but all of these figures reflect the nonmating season. Moreover, in the Takasakiyama troop, as already noted, high-ranking males will take an interest in infants and groom them (Itani 1959).

The information on other species of macaques is less complete. In a captive group of pigtail macaques, *Macaca nemestrina,* observed over a four-year period, social grooming was the single most frequent contact interaction, and females both gave and received more grooming than did males (Bernstein 1972, pp. 404–407). Also, grooming bouts involving females lasted longer than those involving males. Older individuals of both sexes received more grooming than younger ones, with the exception of infants under three months of age. However, older males were more likely to groom than were younger males, whereas young females groomed as frequently as their elders. In both sexes, there was a tendency for the older individuals to engage in longer grooming bouts. In a six-week study of a captive group, adult and subadult males were found to perform less groom-ing than expected, but adult females also performed less than expected

(Jensen et al. 1968). Juvenile males and females, however, both groomed more than expected. Adult males, however, only groomed adult females. There was a tendency, not statistically significant, to groom individuals that ranked higher than oneself. Juveniles and high-ranking females were the most frequent groomers. Subsequent studies of this group in comparison with a bonnet macaque colony formed at the same time has shown the importance of genealogical ties among the pigtails, especially for grooming, where about half is directed within matrilines for adults (Rosenblum 1971). Bertrand's data on *Macaca arctoides (speciosa),* based on a colony of immatures of both sexes and a colony of female immatures, do not show a relationship of grooming to a linear rank hierarchy, but they do show that two of the three male juveniles received about three times more grooming than they gave (Bertrand 1969, chap. 8). Moreover, qualitative observations suggested a rank order effect for males.

The bonnet macaque, *M. radiata,* of South India differs from the other macaque species in important ways. It has been reported to have no sexual asymmetry of grooming frequencies in the field, and both sexes groom each other frequently, though there is a tendency for same-sex grooming. Also grooming among adult males is common, and adult males groom all age and sex categories, but they groom infants and juveniles infrequently (Simonds 1965). Male grooming among bonnets is much higher than among the Japanese macaque, according to Sugiyama, who has observed both species (1971, p. 264). Juveniles and infants groom less than older animals. Mothers, however, groom infants frequently. Grooming among males is reported not to have any relationship to the male dominance hierarchy (Nolte 1955; Simonds 1965; Sugiyama 1971, p. 257). The highest grooming frequency occurs in the months when sexual activity is highest (Rahaman and Parthasarathy 1969, pp. 153–154). In some situations, bonnet troops can have a more fragmented social organization, in which adult males do not have any grooming relationships at all with each other, or in which only a single adult male may be present in the troop. Unlike Japanese, rhesus, or pigtail macaques, captivity studies indicate that grooming among bonnet matriline members falls to chance levels by the end of an individual's first year (Rosenblum 1971, p. 82). However, in bonnet colonies set up with four females and one adult male, females were observed to groom more than males (Rosenblum, Kaufman, and Stynes 1966).

The baboons of the genus *Papio* have at least two different forms of social organization, a unimale troop, consisting of a single breeding male and multiple females, found in *Papio hamadryas* and *Papio papio,* and a multimale troop, with multiple breeding adult males arranged in a dominance hierarchy, as found in *Papio ursinus* (Kummer 1968*b*). These differences in social organization are reflected in the social grooming network. In the unimale troop species, male-female grooming is more likely to be reciprocal, while in the multimale troop, males typically do not groom females except during mating. The baboon multimale troop is similar in this respect to non-*radiata* macaques. In a South African troop of *Papio ursinus,* sexually cycling (nonpregnant and nonlactating) females contributed more grooming than expected, about 57 percent of total grooming to all age and sex categories, though giving most to adult males, other cycling females, and subadult males. Adult males only groomed sexually swollen females, though they received much more grooming than they gave, nearly all of it from females. The grooming distribution was affected by the stage of the estrus cycle, with sexually swollen females engaging in more grooming of adult males and the adult males grooming them, usually in a sexual consort relationship. Adult males gave the shortest grooming bouts. The females groomed other females in the flat part of the sexual cycle. Pregnant females were largely nongroomers (Saayman 1971). Infants were not observed to groom, though they received a small amount from lactating females. The lactating females themselves were groomed no more than expected (Hall 1962; DeVore 1963; Hall and DeVore 1965). In an earlier field study of *Papio ursinus* by K. R. L. Hall, adult females were found to be groomers at about the level that would be expected from their frequency in the troop. However, the females groom for longer periods, about 6.5 minutes compared to 1.3 minutes for the males. Male grooming of females during consort relationships was also observed. Similarly, Bolwig reported that grooming in this species was primarily a female behavior, with male grooming of females being brief (1959, pp. 151–152). Seyfarth found that lactating females were groomed most frequently by other females, but estral status did not affect female grooming (1976). Social rank did have an effect, however. High-ranking females are groomed by a greater number of individuals than low-ranking ones, and grooming went from subordinate to dominant in eighteen of twenty-six pairs of nonlactating females. There was no relationship of duration of bouts to rank. Females groom animals that are in ranks closest to their own. The situation in *Papio anubis* is still not clear. In a small captive group of *Papio anubis,* studied by Rowell, 65 percent of the grooming went from dominant to subordinate, and females groomed more than males (1966). In another colony, an adult and a subadult male were found to groom sexually swollen females more than those in any other reproductive state, and these females groomed males more and females less (Rowell 1968). Higher-ranking females groomed more, with one exception out of five.

In the unimale troop species of *Papio,* adult male grooming of females can occur outside of sexual contexts, and intermale grooming also takes place. In *P. hamadryas* adult males without females often groom each other, but mutual avoidance is practiced by the harem males. Within the harem, however, grooming is the most frequent social activity. Harem adult males perform grooming during 12 percent of observation time and receive it during 19 percent (Kummer 1968a, pp. 44–46, 53, 94). This difference is statistically significant, but the absolute quantity of male grooming is nonetheless high. Grooming by females is concentrated on the male and their own infants. They do little among themselves and compete with each other for an opportunity to groom the male. In this respect, too, this society differs radically from the multimale troop. Young males will follow the unimale troops and they will be groomed by the adult females. When the female is in estrus, the subadult followers may groom the female. In contrast to the multimale troops, the female both receives and performs *less* grooming with her male when she is in estrus. In a study of *Papio papio* at the Brookfield Zoo in Chicago, Boese found that adult males formed one-male subgroups with one to four females each (1975; Dunbar and Nathan 1972). In these subgroups, there was high reciprocal grooming between the adult male and the females. Two forms of grooming were observed: the relaxed grooming frequently described for baboons, and a "rapid grooming" pattern directed to an individual who was being threatened by another. The effect was to remove the groomee from the interaction. If the rapid grooming scores are included in the total, some males groomed more than females. Most grooming, from 100 percent to 67 percent, occurred between subgroup members.

In contrast to the monkey species just considered —indeed in contrast to nearly all other primates —the chimpanzee, *Pan troglodytes,* has elaborated social grooming more among males than among females and the males groom primarily among themselves. In this respect, the grooming groups of adult male chimpanzees resemble the grooming groups of socially immature hamadryas and gelada males who have not yet founded their own unimale troops. Chimpanzees are unusual, however, in the extent to which this male-centric social group persists into adulthood and the extent to which it occurs with a marked diminution of female-to-female grooming. In chimpanzees, more than half of social grooming is between adult males, and there are strong rank-related effects, similar to those recorded for rhesus monkeys. Simpson has described an age-related status hierarchy among males at Gombe Stream Reserve, assessed by wins and losses of spatial supplantations and on pant-grunt vocalizations (1973). The higher a male's status, the longer and more often he is groomed by other males. Also, the higher a male's status, the less long he grooms other males. Males groom more frequently than females, and Sugiyama shows that subordinate-to-dominant grooming is often to placate the dominant animal (1969, pp. 208ff). Other forms of grooming also occur in chimpanzees. Mother-child grooming becomes more important as the child matures, and it becomes more mutual and longer in duration. Also, although the child spends less time with its mother as it matures, a greater proportion of the available time is spent in grooming (Lawick-Goodall 1967). As in rhesus and baboons, adult male-female grooming is likely to mean that the female is in estrus. Some familial female-female grooming occurs, but it is not as elaborated as interfemale grooming in macaques and nonhamadryad baboons. Chimpanzees provide an interesting test case for the postulated motivations underlying grooming because they illustrate the two directions of flow in relatively pure form. Mother-child grooming satisfies the criteria for relaxed familiarity, since it is related to attachment rather than to rank, and the intermale grooming has a large displacement component, as reflected in its upward directionality.

In primate species that are sexually dimorphic for body size, such as the Old World monkeys and the great apes, there is also a systematic relationship between social rank and sex: the males normally outrank the females, where rank is measured by the ratio of wins to losses in aggressive encounters, not by the frequency of aggression. In these species, the greater frequency of female grooming could also be explained as displacement activity created by ambivalence in the lower-ranking age-sex category. In species dimorphic for size, females would be expected to groom more for two reasons. They are the primary care givers to children and they are also more likely to be conflicted when in proximity to males than the males would be when in proximity to them. The baboon and macaque data accord well with this expectation, but the chimpanzee, which is dimorphic for body size but low on female grooming, remains an anomaly. It is clear that additional factors must be at work, and sexual grooming provides a clue.

In all the species considered in this section, symmetrical or near symmetrical grooming occurs in the copulatory context. This is true of the hamadryas baboon as well, even though the total amount of grooming for a female decreases when she is in estrus and the proportion of grooming devoted to the group leader declines (Kummer 1968a, pp. 40–41). In sexual relationships between individuals of different rank, the close relationship between sexuality and agonism can be expected to create special problems. Mutual sexual attraction requires bilateral approach behavior, but simultaneous approach by two parties is not the same social or motivational situation as unilateral approaches by either party. Bilateral approach is not a pair of unilateral approaches, and this conclusion can be inferred from the character of agonistic behavior itself. Among individuals of well-established relative rank, in which the outcome of agonistic interaction can be reliably predicted, the emotional weight of approach-avoidance behavior is borne by the subordinate. This is supported by Rowell's observation of baboons where rank was more predictable from the behavior of the subordinate than of the dominant (1966).[3] Approach toward the dominant is fear-inducing to the subordinate, as is the approach of the dominant toward him. In both kinds of unilateral approach, the social contact requires an inhibition of fear behavior on the part of the subordinate, as revealed by the affiliative forms of submissive facial expressions and gestures. In neither form of unilateral approach must the dominant consider the effect of his own behavior on the other. However, when the dominant wants something from the subordinate, assessment of the effects of his own behavior contribute substantially to his prospective success. In such mutual approaches, both parties are motivated to assess the effects of their own behavior on the partner and to modulate their own behavior on the basis of this feedback from the other. The same logic applies when the dominant animal wants something from the subordinate, regardless of the motivations of the latter. In this case, too, the dominant must assess the consequences of its own actions on the other. Psychologically speaking, a dominant animal with an ulterior motive is equivalent to a subordinate animal. As a consequence of this, a dyadic social interaction between an ulteriorly motivated dominant animal and its subordinate ought to be equivalent to a positive interaction between two animals of indeterminate rank. Ulterior motive in the dominant animal transforms the predictable, rank-related relationship into a process of mutual accommodation. In such cases, the social interaction ought to proceed as a mutual approach-avoidance dance, modulated by facial expression and gestural signals. Like other negotiated relationships already considered, the approach-avoidance gradient should stabilize at the level of mutually rewarding behavior. In species where one sex outranks the other, sexual drive confers ulterior motivation on the dominant member of a male-female pair, and this fact in turn transforms rank-related male-female behavior into a process of mutual accommodation of which the outcome is mutual reward. In reference to grooming behavior, sexuality ought to transform asymmetric displacement grooming of male-female interaction into symmetrical grooming. This is exactly what happens, for symmetrical grooming is typical of the copulatory behavior of macaques, baboons, and to some extent chimpanzees, even though asymmetrical grooming or an absence of male-female grooming is otherwise the norm in these species. Moreover, the consummatory behavior of sexuality, namely copulation, results in a temporary shift of physiological response toward parasympathetic (or "trophotropic") control, as indicated by the relaxed postcopulatory state; and this mood change would facilitate hygienic grooming. Support for this is provided by the fact that orangutans have been seen to groom occasionally after copulation, even though social grooming is not typical in this species (MacKinnon 1974, p. 51).[4] Precopulatory and postcopulatory grooming, although perhaps behaviorally identical, may have different entry pathways: the former being more displacement, the latter more hygienic. Since grooming is itself a relaxing activity, it follows that prolonged sexual behavior can create the conditions for relaxed familiarity and further transform the sexual dyad into a genuine attachment relationship with grooming no longer dependent on the disparity of rank. This may occur in *Theropithecus gelada*. As noted, grooming in this species is nearly all reciprocal, with males grooming their females and females them, and contrary to the other baboons, estrus in the female does not alter the male-female relationship (Dunbar 1978). This suggests that sexual grooming, if it exists as a distinct category at all, has the properties of relaxed familiarity and is unaffected by short-term status changes.

THE GIFT OF GROOMING,
THE REWARDS OF AMBIVALENCE

The hypothesis that nonfamilial symmetrical grooming results from ulterior motives in the dominant animal requires that grooming behavior be both rewarding and accessible to instrumental control. Provided that the initial fear of reduced social distance can be overcome, there is good behavioral evidence that grooming is rewarding. Groomed animals not only convey enjoyment by their posture and expression, but animals will solicit grooming from others. Tickling, too, which may be related, is used often as a reward in home-reared apes.[5] Moreover, a chimpanzee has been taught new problem-solving tasks with grooming as its reward (Falk 1958). In the monkey and ape species that socially groom, there are specific gestures and postures that solicit grooming. Among the stumptail macaques, *Macaca arctoides (speciosa)*, studied by Bertrand, grooming and solicitation were the primary means of establishing contact with strangers or of renewing contact with a former member of the group (Bertrand 1969, chap. 8). In these cases, the perineal present, which resembles a sexual present, was the most frequent form of solicit. In all colonies studied, grooming peaked soon after the establishment of the group, fell to baseline, and then would increase as new individuals were introduced. The highest-ranking individual would groom a dominant newcomer, including humans. Females also used grooming to get access to another's baby, starting usually at the mother's back and working toward the infant. Appeasement grooming, after an altercation, and displacement grooming during excitation, also occur. Also, grooming by the aggressor occurred in bullying situations and the recipient "froze." In this species, there are also a number of stereotyped groom solicits: head presenting, neck presenting, chest presenting, chest presenting with one arm up, lying down, perineal presenting, and a short pregroom of a few seconds. The potential groomer may walk away, ignore, request grooming or do something else. Mothers will groom infants for long periods of time, and stumptails will often groom pieces of fur or stuffed animals. They will also groom the wounds of injured animals, even if the latter are not ordinarily grooming recipients. In another instance of apparent instrumental grooming, an orphaned female infant began to groom much more frequently after her mother died, and she used it to gain access to the biggest male. When she was placed in another group where a juvenile protected her, she reverted to a low grooming frequency. The phenomena that Bertrand observed in stumptail macaques have been widely reported for other socially grooming species. Nonhuman primates appear to use grooming instrumentally to get what they want; they find grooming rewarding and have standardized gestures for soliciting it from others; they give behavioral indications that grooming is an innate action with an intrinsic motivation of its own; and they also direct grooming to suitable body parts and foreign bodies, suggesting that it is a goal-directed hygienic activity. These generalizations are supported by observations in other species as well.

Among hamadryas baboons, the grooming response can be elicited when an animal places itself sideways on all fours in front of its partner, lowers its head, and arches its tail (Kummer 1968a, pp. 44–45). Presenting the rear can also elicit grooming, and animals have been observed pulling on the other to groom him. Grooming also appears to be used instrumentally. An old male never groomed his females when he had seven of them, but he began to groom them when his harem was reduced to two. Also, males are more solicitous with their first females than with later additions. In *Theropithecus gelada,* the chest present and rear present are also used to solicit grooming, and perhaps sexual contact as well (Dunbar and Dunbar 1975, pp. 34, 81). In an interesting observation, the Dunbars report that when a new female entered a unimale troop, she groomed the highest-ranking female (as measured by the initiation of troop movements), but the highest-ranking female did not groom her. In Sykes monkeys, *Cercopithecus mitis kolbi,* studied in captivity by Rowell, grooming was the most frequent form of interaction. It could be solicited by lying down with the back toward the other or by offering the nape of the neck (1971, pp. 627–628). The solicits found in macaques and baboons also occur but less frequently. The New World monkey, *Cebus nigrivittatus,* solicits by thrusting out its chest or by lying down with back or side presented (Oppenheimer and Oppenheimer 1973, p. 414). Grooming is also solicited by the presentation of body parts in *Colobus badius, Gorilla gorilla beringei,* and other species of *Macaca.* The prosimian, *Lemur fulvus,* also will solicit by presenting the ventral surface of the forearm or head (Harrington 1975, p. 464).

Sparks points out that groom solicits usually involve postures and gestures that require the avoidance of eye contact. As such, groom solicits are similar to appeasement gestures, and in the perineal present and supine posture, the two classes of gesture overlap (1967, p. 152). An exceptional groom solicit was observed by Struhsaker among vervet monkeys, *Cercopithecus aethiops,* where the potential recipient would sit in front of the potential groomer and hold the latter's face with its hands for about two seconds, although this may be derived from nuzzling, described later (1967, p. 5). However, the presentation of body parts to be groomed is also the normal solicit in this species as well. Given that contact between individuals of different ranks is potentially aversive, the relationship of groom solicits to appeasement gestures is not surprising. However, primate grooming behavior is closely integrated with the capacity for volitional action and intelligent interpretation of the acts of others. A groom solicit can evoke a range of possible responses. Sometimes the groom is given, but at other times an ignore or a counterrequest is evoked. For groom solicits, Rowell reports a one-third success rate among baboons and a 56 percent success rate in Sykes monkeys (1966, p. 438). Although sometimes called social releasers, solicits do not conform to the definition of a releaser because their effect is highly dependent upon contextual considerations, whereas classical releasers are basically context-free.

Nonetheless, there is now some question as to whether solicits are really soliciting. In Japanese macaques *(Macaca fuscata)*, the only species so far analyzed in this way, grooming interactions among adult females are usually preceded by a *vocal* request by the would-be groomer (Mori 1975, p. 115). Like groom solicits, the response was variable, with grooming taking place in about a third of the cases. However, this study revealed that the behavior patterns that have generally been termed *groom solicits* —lying on one side, lowering the head, and lying on the stomach—were usually groom *acceptances,* given by the solicited party if it accepted the vocal request. This study confirmed, however, that grooming usually takes place through a process of mutual volitional acts, and Mori has schematized Japanese macaque grooming by the following interactive sequence: (1) approach of groomer to groomee, (2) groomer utters groom vocalization, (3) groomee gives groom solicit gesture, (4) groomer utters groom vocalization, (5) groomer approaches, (6) groomer grooms groomee. These interactions, although similar to the ritualized exchanges described by classical ethology for nonmammalian vertebrates, differ from the latter in one important respect: the behavior is under volitional control and can be broken off by either party.

Volition presupposes the ability to inhibit behavioral responses. This capacity has generally been recognized by ethologists by their inclusion of the behavior pattern termed *ignore* in a wide spectrum of primate ethograms. Behaviorally, an ignore is manifested by gaze direction toward the act of another followed by gaze steadfastly turned away. In rhesus monkeys the lookaway often persists through tactile contact by the soliciting party. As already noted, groom solicits are often ignored. Second, volition entails the ability to assess the external consequences of action. In this respect, it is very different from the instincts described by ethologists where the goal of the action is the release of the consummatory behavior, quite irrespective of instrumental efficacy. Third, volition presupposes the ability to utilize a behavior pattern instrumentally to achieve previously assessed effects. Although none of these properties logically presuppose learning, volitional behavior is empirically associated with learning capacity in higher primates. As such, a fourth characteristic is the capacity to compare the actual consequences of action with anticipated consequences of action and to modify subsequent performances on the basis of this experience. When social interactions distribute rewards through mutual, volitional behavior, there emerges a class of interactions best termed *prestational*.

Prestational interaction can be distinguished from the other forms of interaction already considered by the conjunction of several characteristics. First, in prestation, the interaction is mediated through behaviors that function as *offerings, requests, acceptances,* and *refusals.* Implicit in these distinctions is the possibility of refusing or accepting or offering or withholding, and this possibility requires the volitional control of behavior. Consequently, early social science formulations, such as that of Mauss, which focused on the obligatory nature of gift giving in human societies, confused the normative control of prestation interaction with the prestation process itself (Befu 1977; Sahlins 1965). Anthropologists have subsequently shown that refusals do occur and can be systematically used to alter social relationships (Young forthcoming). Other social scientific approaches, in attempting to incorporate coercion and altruism into a prestational framework, have overextended the concept to areas where it is inappropriate. Ethological studies indicate that altruistic interactions are not primarily volitional but mediated by innate attachment mechanisms. Maternal grooming, for example, is primarily a one-way interaction, and it has been necessary to develop a special body of genetic theory, namely inclusive fitness, to account for this behavior within a Darwinian framework (Trivers 1971). Nonetheless, it has been shown that altruistic behavioral mechanisms, in which the individual makes no assessment of the benefit to himself but simply behaves selflessly, are compatible with evolutionary theory provided that more genes are passed on, via his relatives, than would be by an act of selfishness. Since interactions mediated by altruistic mechanisms are not necessarily volitional, they give no scope to offering, requesting, or refusing and cannot be considered prestations. A similar caveat applies to coercion. The instrumental use of aggression to effect exchange negates the possibility of refusal, so that such forced interactions are not prestations. For example, in one group of *Propithecus verreauxi,* one adult male named F frequently approached the other male P, thrust P's head into contact with his (F's) shoulder, and forced P to groom him, which P did with submissive gestures (Richard and Heimbuch 1975, p. 318). Operationally, coercion can be recognized by fearful signals by the donor.

The second characteristic of prestation interactions is that a reward must be given to one of the partners by the other as a consequence of the interaction. Affiliative interactions in which no rewards are exchanged by overt behavioral acts are excluded by this definition, as is Homans's usage of punishments being exchanged (Befu 1977, p. 258). Ethology would regard the exchange of punishments as agonistic activity. Prestation, as used here, is far narrower than the exchange concept as used in social science, but some primate grooming interactions nonetheless satisfy the criteria of the definition. As already argued, grooming must be seen as a reward, since it can be used as a reinforcer to acquire overt behavior, and many species of socially grooming primates treat it as if it were rewarding. Grooming is also frequently given in a volitional context in which it can be offered, solicited, accepted, and refused. The emphasis that contemporary *Homo sapiens* places on material objects has obscured the evolutionary implications of grooming behavior. Many species of nonhuman primates have grooming networks that contain both altruistic and self-interested forms of grooming; the self-interested forms of grooming are formally similar to prestational interactions that occur in man; grooming can also be coercively induced; and it bears systematic relationships to social rank on the one hand and to familiarity on the other. In all of these properties, grooming is similar to the networks of object exchange that anthropologists have described for human societies. These facts suggest that the grooming networks of nonhuman primates are ancestral to the object-exchange networks of modern man. This hypothesis is supported by the nature of grooming itself.

FROM MOUTH TO HAND

Although the discussion has hitherto treated grooming as if it were simply an invariant behavior pattern, a fine-grained analysis of grooming shows it to be a motor program of some complexity. (1) It involves the most dextrous manipulations of peripheral effectors a species is capable of, including the thumb-index finger opposition of the Old World monkeys, which Napier termed the *precision grip* (1962). (2) It requires organized sequences of behaviors, including holding the fur, parting it, and picking out fine particles. (3) It is goal-directed by objects in the external world at several levels of control: by the choice of the individuals groomed, by the part of the body selected, and by the particular foreign bodies discovered there. (4) It frequently requires functional differentiation between the hands, with one hand parting the fur and the other grasping. (5) It often involves the integration of distinct grooming actions. For example, Horwich has described three distinct hand postures and motions used by spectacle langurs *(Presbytis obscura)* in grooming their infants (1974, pp. 153–154), and Sade has shown the complexity of the rhesus grooming repertory (1972). As noted earlier, behavior patterns can often be executed in decerebrate mammals, but Rioch, in an early summary of this literature, specifically pointed out that grooming behavior in decerebrate cats was conspicuous by its absence. In the primate, too, there is no doubt that grooming is a complex motor task that involves the phylogenetically advanced mechanisms of sensory and motor control. Grooming, in other words, is already a program for the dextrous manipulation of objects. Consequently, as the innate capacity for object manipulation becomes more sophisticated in the course of human evolution, then other object-manipulation programs ought to co-occur with the more traditional grooming behavior. The comparative evidence supports the interpretation of progressive motor elaboration, and it is reflected first of all in the *distalward* progression of skilled effectors. The prosimians socially groom with their mouths, and they use both their mouths and toilet claws on themselves (Szalay and Seligsohn 1977). Monkeys and apes socially groom with their hands (with some partial exceptions such as langurs, which have reduced thumbs), but apes have been observed to self-groom with objects such as sticks and crumpled leaves (MacKinnon 1974, p. 51).[6] *Cebus* monkeys can do this as well. If social behavior becomes mediated by more distal effectors as primate evolution proceeds, then it is a logical next step for hominids to abandon to some extent direct behavior and operate upon intermediate objects instead. In this model, as object-manipulation programs become more sophisticated, their use will lead to the exchange of rewarding objects where apes exchange rewarding behavior.

If object-using programs are incorporated into grooming behavior, then actual object exchange should follow grooming networks. Some support for this is provided by studies of predation among chimpanzees (Teleki 1973, pp. 146–150, 158–159, 162–163). When chimpanzees have killed an animal, there is some distribution of the carcass. In 616 instances of distribution reported on by Teleki, 37 were examples of recovering meat that had fallen to the ground; in 184 cases the meat was taken; and 395 instances were classified as requests. The requests were performed by (1) peering intently at the face or meat of the other, (2) touching the face or meat of the eater, (3) extending the hand palm upward, and (4) emitting *hoo* or *whimper* vocalizations either alone or in conjunction. Of the 395 requests, only 144 were successful. However, voluntary handing over was only observed four times. The age-sex categories involved in redistribution provide support for the relationship between object exchange and grooming networks. Since adult males do the hunting, they also control the carcass. Consequently, adult males are the target for requests by both males and females, and adults are overwhelmingly represented in the redistribution process. An exception is the matrifocal family, where children often receive meat from their mother, even though the meat does not usually pass through more than one individual. Also, in 132 instances of redistribution involving an adult female with an adult male, estrus females had a success rate of 69 percent compared to 40 percent for nonestrus females. Estrus females are not only more likely to be in proximity to adult males but are also more persistent in requesting meat. In a much earlier study by Yerkes, with artificially paired male and female chimpanzees in the laboratory, females also obtained more male controlled resources when in estrus. Also, since grooming among males is rank-related, high-ranking individuals should obtain more meat on redistribution. This appears to be the case. The alpha male Mike consistently got access to choice pieces and apparently ate more, as did Flo, the highest ranking female. Moreover, the report that aggression is low during redistribution is consistent with the grooming hypothesis, for grooming is not distributed by aggression either. These data at least do not contradict the hypothesized relationship between grooming network and object exchange network, and the differential distribution from males to estrus females and from mothers to dependent children is a close correspondence.

Some further corroboration is provided by other cases of object transfer among primates. Sykes monkeys *(Cercopithecus mitis kolbi)*, studied in captivity, often took or attempted to take food from each other (Rowell 1971, pp. 632–633). As Rowell remarks, the behavior bore more similarity to friendly behavior than to aggression. If a social hierarchy is computed on the basis of retreating behaviors, then 60 percent of food-taking attempts were directed up the hierarchy, about half of which were successful. Of the down-directed behaviors, 68 percent were successful. Groom solicits were also directed both up and down the hierarchy, about 50 percent in each direction, with about a 50 percent success rate. Also, in a captive group of gibbons, individuals often took food from one another, but the food passed in both directions, not asymmetrically as would be expected in agonistically mediated object exchange (Berkson 1966; Berkson and Schusterman 1964). Among a large zoo colony of *Pygathrix nemaeus,* on four occasions monkeys were observed to hand food (branches of the shrub *Eugenia*) to others, even though no solicitation was observed, and the transfers did not follow rank-related paths (Kavanagh 1972).

The signals that accompany food transfer are also incompatible with aggressive mediation and often overlap with the gestures that occur in grooming contexts. In Sykes monkeys the food transfer is often preceded by a nuzzle gesture, in which the food-receiver monkey noses against the mouth of the food-holding monkey. This gesture is also used as a greeting, and it is morphologically similar to the muzzle–muzzle gesture described for a closely related species, the vervet monkey, *Cercopithecus aethiops* (Struhsaker 1967, p. 44). In both species it is primarily directed up the hierarchy, particularly to the highest-ranking male, and it is directed primarily to feeding animals. In vervets, the muzzle–muzzle gesture is present in infants at least by fifty-four days, and it also occurs in grooming and after intense agonistic encounters. In the begging of chimpanzees, frequent stratagems are to peer intently while thrusting the mouth very close to the face or meat of the feeding animal, or to reach out and touch either the meat or the lips of the other (Teleki 1973, p. 148). All of these actions are probably intention movements of feeding, rather than agonistic. In savanna baboons, where food transfer is predominantly agonistically mediated and restricted to cases where prey has been killed (discussed later), intention movements of feeding nonetheless occur in grooming contexts.

A common gesture used in grooming in *Papio* is lip smacking. As Anthoney points out, it is not a lip smack at all, but a click produced when the "moistened anterior dorsum of the tongue is laid flat against the backs of the incisors and anteriormost hard palate; then the contact between the lingual and dentopalatal surfaces is repeatedly broken and reestablished by pulling the posterior part of the tongue rapidly back and forth (six to eight times per second)" (1968, p. 360). Anthoney has interpreted this behavior as homologous to the nursing movements of the infant, whereas Redican derives it from the mouthing of small particles found during grooming. Since Redican confirms Anthoney's description of the movements involved in the acoustic component, it is hard to see what role these movements have in grooming (Redican 1975, p. 138). Anthoney's interpretation is given additional support by the fact that lip smacking generally appears at a very early age in primates. The lip-smacking component has been described as present at three days of age in rhesus monkeys, whereas grooming appears much later. However, in baboons, some of the finger movements of grooming can accompany nursing in the neonate. In baboons, the use of lip smacking is often in mutual interactions. Adult females lip smack to infants while grooming their mothers and locomoting infants may lip smack to females who lip smack to them (Anthoney 1968; Rowell et al. 1968, pp. 471–472). Adult males often lip smack to exploring infants as well. Lip smacking is also part of the copulatory behavior of the female baboon, who looks over her shoulder and lip smacks to the mounting male.

The relationship of lip smacking to rank has been variously reported in primates, with Rowell finding that 85 percent of lip smacks in baboons were from dominant to subordinate, whereas authors working with other species have implicated it in subordination (Rowell 1966, p. 438). In an episode described by Hall, when a captive male baboon was released near a wild troop, lip smacks were an important part of the greeting behavior, but the vast majority of greeting approaches were made by animals the same size or smaller than the released male, confusing the issue of rank (1962). Moreover, there was a sex difference in the greeting behavior. Females and the released male tended to kiss on the mouth, whereas the males tended to lip smack. However, lip smacks were common and they were often mutual. In all the lip smacking primates studied, lip smacking is closely associated with grooming, greeting, and sexuality. Moreover, Anthoney has noted that in baboons "greeting and sexual behaviors are so closely related ontogenetically that they are best treated together" (1968, p. 359). Grooming, greeting among near-equals, sexuality, and adult–infant relations are all contexts of mutuality where rank relationships are least relevant. The morphological similarity of the behavior to both nursing and paragrooming mouth movements indicate that it is systematically related to the negation of rank rather than to rank itself. The intention movements of ingestion are interpreted as rank-reducing or affiliative gestures, even though no actual foodstuffs are transferred. In baboons, however, the actual consummatory acts of ingestion do occur, because lip smacking is often the precursor to mouthing of the genitals, mouth, or nipples, even in adult interactions (Hall 1962; Anthoney 1968).

The baboons also provide an apparent exception to the hypothesis that object transfer follows grooming networks. Hunting by baboons, with subsequent meat transfer, has been reported from all over Africa. For example, *Papio cynocephalus* in Kenya have been observed to hunt a variety of animals including vervet monkeys, hares, young gazelles, rodents, birds, lizards, and frogs (Hausfater 1976; Harding 1975). Unlike chimpanzees, no voluntary or solicited food transfers were observed during the forty-five recorded episodes, and no begging gestures were seen. Like chimpanzees, however, one individual, usually an adult male, generally dominated a carcass, but unlike the apes, there were few examples of more than one animal feeding at a carcass at a time. The baboons also differ from chimps in that agonistic interactions during the postpredation period are high rather than low. Some food transfer does take place, however, by supplanting another at the carcass or by feeding on scraps abandoned by the carcass feeder. In actual redistribution rates, the baboons and chimpanzees are equivalent, but the mechanisms of redistribution are primarily agonistic in the baboon case and affiliative with apes. However, these observations do not contradict the grooming hypothesis but largely support it. The adult males are the major carcass feeders in baboons, and they are groomed by females, but they do not usually reciprocate, except when females are in estrus. Also, adult male baboons, unlike chimpanzees, do not groom each other. Consequently, object transfer would not be expected to follow grooming paths where no such paths exist. The baboons therefore provide an interesting test case. An implication of this reasoning is that such transfer should occur where grooming paths do exist among baboons. This prediction is confirmed among *Papio anubis* in Kenya, studied by Strum (1975). Here infants who had close attachments with their mothers or relationships with the adult male who made the kill had high scores for meat acquisition. Moreover, mutual feeding at carcasses is observed between mothers and offspring and between adult male and adult female pairs. The latter was sometimes a sexual consort of the male. In this group, too, agonistic supplantations at the carcass were typical.

The baboon meat transfer indicates that agonism can indeed be systematically involved in the distribution of resources, as the early "dominance" model argued, but the other evidence already discussed shows that among familiar individuals, the thresholds of consummatory agonism are usually too high to take precedence over the affiliative channel. Some support for this is provided by the interactions of chimpanzees with baboons at Gombe Stream Reserve, where chimpanzees will aggress against baboons to steal the prey they have killed, although they do not usually do this to fellow chimps (Morris and Goodall 1977). Moreover, chimpanzees at Gombe will sometimes kill the baboons, while at other times they will engage in play and grooming. However, what is not deducible from the published accounts is the extent to which the individuals who kill baboons are the same individuals who previously played with them. In fact at the periphery of the social universe in primates, such inconsistencies are not uncommon. In the intergroup encounters of vervet monkeys recorded by Struhsaker or of the rhesus monkeys studied by Lindburg and Southwick, violent agonism and peaceful intergroup interactions can both occur, but the individuals involved may be different and the situation can change rapidly for the worst, indicating the familiarity gradient to be nonetheless important.

Various lines of evidence indicate that social grooming in nonhuman primates is to be regarded as the evolutionary precursor of hominid economies. The grooming networks of nonhuman primates can be transformed into object transfer networks by substituting advanced object-manipulation programs for the simpler object-manipulation program of social grooming itself. Moreover, the elaboration of object-manipulation is generally acknowledged to have occurred in the transition from apes to hominids and to be of major theoretical importance. This hypothesis is supported by the occurrence of gestures derived from feeding in the grooming and affiliative behavior of primates and by the networks of actual object transfer in species that have it. Moreover, object transfer among individuals who do not have grooming relations that flow in the right direction, as among baboons, have only agonistically mediated object transfer in these circumstances. In addition, the morphology of primate grooming and human exchange show important similarities in their relation to rank, familiarity, kinship, and volition. This evolutionary hypothesis has important implications for social science theory. First, it indicates that the emergence of object-transfer in hominids is not to be regarded as the source of economic relations but only as marking the transition to a particularly human form of such relations. Second, economic relations are to be defined, in a phylogenetic sense, as the systematic social flow of rewards, measured by their reinforcement value, regardless of whether they involve the transfer of material objects. By implication, economic relations begin with the volitional control of innate rewards mediated by innate behaviors and are not dependent upon learned behavior per se. Third, the common social scientific goal of formulating a unitary theory of exchange is probably impossible, since grooming itself appears to participate in at least three distinct systems of causality: coercive transfer, altruistic attachment, and prestational interactions. Fourth, by implication from the previous proposition, it is equally incorrect to regard one form of exchange as primary and to derive the others from it. All primates have a care-giving relationship between adult and child that can serve as an altruistic nexus for the flow of rewards, and probably all primates are capable of aggressive acquisition of rewards at the margins of social familiarity. Prestational interactions, although especially elaborated among hominids, nonetheless occur in some New World monkeys *(Cebus),* many Old World monkeys and apes, and the group-living lemurs of Madagascar. Prestational interactions cannot therefore be regarded as uniquely linked to the hominid emergence. On the contrary, their wide geographic distribution indicates that they have either evolved at least three times within the primate order or are derived from a homologous psychological substrate that antedates the dispersal of the primates to their modern ranges, in Oligocene times, over thirty million years ago.

SUMMARY

The grooming behavior of nonhuman primates participates in networks of social exchange that share many properties with networks of material object exchange in man. Among these are systematic differences in grooming in relation to sex, social rank, age, and social group boundaries. Also, grooming functions in prestational interactions characterized by offering, acceptance, refusal, and reciprocation. An evolutionary hypothesis is presented linking human object exchange networks to the supplementation of the instrumental action of grooming by advanced programs of object manipulation and construction. In this way, the object-using programs of hominids become subject to the same social dynamics that characterize the grooming networks of nonhuman primates. Thus human object exchange and nonhuman primate grooming are affectively comparable even though they are behaviorally dissimilar. This perspective requires that wealth be defined as the tokens used in exchange relationships, and human wealth differs from the wealth of monkeys in that in man the tokens are material object manipulations socially directed rather than direct manipulation of a social object. This postulated change is in accord with the distalward migration of instrumentality and the affective control of instrumentality previously discussed. The evolutionary transformation of grooming networks into object-exchange networks makes possible an accumulation of wealth, which is not possible to monkeys. Consequently, it is the accumulation of human exchange tokens (and the social consequences that follow from that), not acquisitiveness per se, that is the emergent feature of hominid society. This evolutionary model makes exchange behavior, not property, the phyletic precursor of exchange networks in man, because exchange is fundamentally a process of social interaction. From this perspective, exchange emerges before the incorporation of material objects into exchange relationships, and the first wealth is behavioral, not material. The aggressive mediation of object transfer also occurs in nonhuman primate societies, but it also co-occurs with object transfer that follows grooming relationships. However, models that regard the aggressive control of property as the precursor of hominid exchange ignore the fact that prestational behavior — which includes offering, accepting, and refusing — occurs not in primate aggressive encounters but primarily in grooming encounters.

NOTES TO CHAPTER 5

1. This has also been confirmed by Mitchell and Tokunaga (1976).

2. However, Eisenberg and Kuehn (1966, pp. 31ff) give evidence for interadult grooming, but the highest ranking adults groom most frequently, suggesting that it too is care-giver grooming (see following).

3. This does not imply, however, that rank itself is only a product of the subordinate's behavior.

4. For effects of sex on grooming generally see Merrick (1977) and Crawford (1940).

5. Tickling shares with grooming a rewarding mood change induced by tactile stimulation by another, although the mood is different.

6. See also chap. 4 of this book.

REFERENCES

ALEXANDER, B. K., and BOWERS, J. M.   1969. Social organization of a troop of Japanese monkeys in a two-acre enclosure. *Folia Primatol.* 10:230–242.

ANDREW, R. J.   1956. Normal and irrelevant toilet behaviour in *Emberiza* spp. *Brit. J. Animal Behaviour* 4:85–91.

ANTHONEY, T. R.   1968. The ontogeny of greeting, grooming, and sexual motor patterns in captive baboons (superspecies *Papio synocephalus*). *Behaviour* 31:358–372.

BEFU, HARUMI   1977. Social exchange. *Ann. Rev. Anthropology* 6:255–281.

BERKSON, G.   1966. Development of an infant in a captive gibbon group. *J. Genet. Psychol.* 108:311–325.

BERKSON, G. and SCHUSTERMAN, R. J.   1964. Reciprocal food sharing of gibbons, *Primates* 5 (1–2):1–10.

BERNSTEIN, I. S.   1968. The lutong of Kuala Selangor, *Behaviour* 32:1–16.

1972. Daily activity cycles and weather influences on a pigtail monkey group. *Folia Primatol.* 18:390–415.

BERNSTEIN, I. S. and MASON, W. A.   1963. Activity patterns of rhesus monkeys in a social group. *Animal Behaviour* 11:455–460.

BERNTSON, G. G. and HUGHES, H. C.   1974. Medullary mechanisms for eating and grooming behaviors in the cat. *Experimental Neurology* 44:255–265.

BERTRAND, MIREILLE   1969. *The behavioral repertoire of the stumptail macaque.* Basel: Karger.

BOESE, GILBERT K.   1975. Social behavior and ecological considerations of West African baboons *(Papio papio)*. In *Socioecology and psychology of primates*, ed. R. Tuttle, pp. 205–230. The Hague: Mouton.

BOLWIG, NIELS   1959. A study of the behaviour of the chacma baboon, *Papio ursinis*. *Behaviour* 14:136–163.

CARPENTER, C. R.   1935. Behavior of red spider monkeys in Panama. *J. Mammalogy* 16:171–180. Reprinted in *Naturalistic behavior of nonhuman primates*, ed. C. R. Carpenter. University Park: Pennsylvania State U. Press.

CHALMERS, N. R.   1968. The social behavior of free living mangabeys in Uganda. *Folia Primatol.* 8:263–281.

CHALMERS, N. R., and ROWELL, T. E. 1971. Behaviour and female reproductive cycles in a captive group of mangabeys. *Folia Primatol.* 14:1–14.

CHIVERS, DAVID J. 1972. The siamang and the gibbon in the Malay Peninsula. *Gibbon and Siamang,* vol. 1, pp. 103–135. Basel: Karger.

1974. The siamang in Malaya. A field study of a primate in tropical rain forest. *Contributions to Primatology,* vol. 4. Basel: Karger.

CRAWFORD, M. P. 1940. The relation between social dominance and the menstrual cycle in female chimpanzees. *J. Comp. Psychol.* 30:483–513.

DELIUS, J. D. 1970. Irrelevant behavior, information processing, and arousal homeostasis. *Psych. Forsch.* 33:165–188.

DELL, P.; BONVALLET, M.; and HUGELIN, H. 1961. Mechanisms of reticular deactivation. In *Ciba Foundation Symposium on the nature of sleep,* ed. G.E.W. Wolstenholme and Maeve O'Connor, pp. 86–102. London: Churchill.

DEVORE, IRVEN 1963. Mother-infant relations in free-ranging baboons. In *Maternal Behavior in mammals,* ed. H. Rheingold, pp. 305–335. New York: Wiley.

DOYLE, G. A.; ANDERSON, A.; and BEARDER, S. K. 1969. Maternal behaviour in the lesser bush-baby (*Galago senegalensis moholi*) under semi-natural conditions. *Folia Primatol.* 11:215–238.

DOYLE, G. A.; PELLETIER, A.; and BEKKER, T. 1967. Courtship, mating, and parturition in the lesser bushbaby *(Galago senegalensis moholi)* under semi-natural conditions. *Folia Primatol.* 7:169–197.

DRICKAMER, L. C. 1976. Quantitative observations of grooming behavior in free-ranging *Macaca mulatta. Primates* 17:323–335.

DUNBAR, R. I. M. 1974. Observations on the ecology and social organization of the green monkey, *Cercopithecus sabaeus,* in Senegal. *Primates* 15:341–350.

1978. Sexual behaviour and social relationships among gelada baboons. *Animal Behaviour* 26:167–178.

DUNBAR, ROBIN, and DUNBAR, PATSY 1975. Social dynamics of gelada baboons. *Contributions to primatology,* vol. 6. Basel: Karger.

DUNBAR, R. I. M., and NATHAN, M. F. 1972. Social organization of the guinea baboon, *Papio papio. Folia Primatol.* 17:321–334.

EISENBERG, J. F., and KUEHN, R. E. 1966. The behavior of *Ateles geoffroyi* and related species. *Smithson. Misc. Coll.* 151, no. 8: 1–63.

FALK, J. L. 1958. The grooming behavior of the chimpanzee as a reinforcer. *J. Experimental Analysis of Behavior* 1:83–85.

GARTLAN, J. S. 1969. Sexual and maternal behaviour of the vervet monkey, *Cercopithecus aethiops. J. Reproduction and Fertility.* Supplement 6: Biology of reproduction in mammals, ed. J. S. Perry and J. W. Rowlands, pp. 137–150. Oxford: Blackwell.

GAUTIER-HION, A. 1971. Repertoire comportemental du talapoin *(Miopithecus talapoin)*. *Rev. Biol. Gabon.* 7:295−391.

HALL, K. R. L. 1962. The sexual agonistic and derived social behaviour patterns of the wild chacma baboon, *Papio ursinus. Proc. Zool. Soc. Lond.* 139:283−327.

HALL, K.R.L. and I. DeVORE 1965. Baboon social behavior. In *Primate behavior: field studies of monkeys and apes*, ed. I. DeVore. pp. 53−110. New York: Holt, Rinehart and Winston.

HANSEN, ERNST 1966. The development of maternal and infant behavior in the rhesus monkey. *Behaviour* 27:107−149.

HARDING, ROBERT S. O. 1975. Meat-eating and hunting in baboons. In *Socioecology and psychology of primates,* ed. R. Tuttle, pp. 245−257. The Hague: Mouton.

HARRINGTON, JONATHAN E. 1975. Field observations of social behavior of *Lemur fulvus fulvus* E. Geoffroy 1812. In *Lemur Biology,* ed. Ian Tattersall and R. W. Sussman, pp. 259−279. New York and London: Plenum.

HAUSFATER, GLEN 1975. Dominance and reproduction in baboons *(Papio cynocephalus).* a quantitative analysis. *Contributions to Primatology,* vol. 7, Basel: Karger.

1976. Predatory behavior of yellow baboons. *Behaviour* 56:44−68.

HINDE, R. A.; ROWELL, T. E.; and SPENCER-BOOTH, Y. 1964. Behaviour of so-cially living rhesus monkeys in their first six months. *Proc. Zool. Soc. Lond.* 143, pt. 4: 609−649.

HORWICH, R. H. 1974. Development of behaviors in a male spectacle langur *(Presbytis obscurus).* *Primates* 15:151−178.

HORWICH, R. H., and WURMAN, CANDIDA 1978. Socio-maternal behaviors in response to an infant birth in *Colobus guereza. Primates* 19:693−713.

HUTCHINS, M., and BARASH, D. P. 1976. Grooming in primates: implications for its utilitarian function. *Primates* 17:145−150.

IERSEL, J. J. A. van, and BOL, A. C. A. 1958. Preening of two tern species: a study of displacement activities. *Behaviour* 13:1−88.

ITANI, J. 1959. Paternal care in the wild Japanese monkey, *Macaca fuscata fuscata. Primates* 2:61−93.

JAY, PHYLLIS C. 1962. Aspects of maternal behavior among langurs. *Annals N.Y. Acad. Sciences* 102, art. 2:468−476.

1965. The common langur of North India. In *Primate behavior: field studies of monkeys and apes,* ed. I. DeVore, pp. 197−249. New York: Holt, Rinehart and Winston.

JENSEN, G. D.; TOKUDA, K.; BOBBITT, R. A.; and SIMONS, R. C. 1968. Interactive relationships of age-sex subgroups in a captive group of pigtailed monkeys *(M. hemestrina). Proc. VIII int congr. anthrop. & ethnolog. sci.* 1:252−254.

JOLLY, ALISON 1966. *Lemur behavior: a Madagascar field study.* Chicago: U. of Chicago Press.

KAUFMANN, JOHN H. 1967. Social relations of adult males in a free-ranging band of rhesus monkeys. In *Social communication among primates,* ed. S. Altmann, pp. 73–98. Chicago: U. of Chicago Press.

KAVANAGH, M. 1972. Food-sharing behaviour within a group of douc monkeys *(Pygathrix nemaeus nemaeus). Nature* 239 *(London):* 406–407.

KENDON, ADAM, and FERBER, ANDREW 1973. A description of some human greetings. In *Comparative ecology and behaviour of primates,* ed. R. P. Michael and J. H. Crook, pp. 591–668. New York and London: Academic Press.

KINZEY, W. G., et al. 1977. A preliminary field investigation of the yellow handed Titi monkey, *Callicebus torquatus torquatus,* in northern Peru. *Primates* 18:159–181.

KLECK, R. 1970. Interaction distance and non-verbal agreeing responses. *Brit. J. Social and Clinical Psychol.* 9:180–182.

KLOPFER, P. H., and KLOPFER, M. S. 1970. Patterns of maternal care in lemurs: I. Normative description. *Z. Tierpsychol.* 27:984–996.

KUMMER, HANS 1968a. *Social organization of hamadryas baboons: a field study.* Basel: Karger.

1968b. Two variations in the social organization of baboons. In *Primates: studies in adaptation and variability,* ed. P. Jay, pp. 293–312. New York: Holt, Rinehart and Winston.

LANCASTER, JANE B. 1971. Play-mothering: the relations between juvenile females and young infants among free-ranging vervet monkeys. *Folia Primatol.* 15:161–182.

LAWICK-GOODALL, JANE van 1967. Mother-offspring relationships in free-ranging chimpanzees. In *Primate ethology,* ed. D. Morris, pp. 287-346. Chicago: Aldine.

LESKES, ANDREA, and ACHESON, N. H. 1971. Social organization of a free-ranging troop of black and white colobus monkeys *(Colobus abyssinicus).* In *Proc. 3rd. int. cong. primatol.,* ed. H. Kummer, vol. 3, pp. 22–31. Basel: Karger.

LINDBURG, D. G. 1971. The rhesus monkey in North India: an ecological and behavioral study. In *Primate behavior: developments in field and laboratory research,* ed. Leonard Rosenblum, vol. 2, pp. 1–106. New York: Academic Press.

LINDBURG, DONALD G. 1973. Grooming behavior as a regulator of social interactions in rhesus monkeys. In *Behavioral regulators of behavior in primates,* ed. C. R. Carpenter, pp. 124–148. Lewisburg, Pa.: Bucknell U. Press.

LOY, J. 1970. Behavioral responses of free-ranging rhesus monkeys to food shortage. *Amer. J. Phys. Anthropol.,* n.s. 33:263–271.

LOY, JAMES, and LOY, KENT 1974. Behavior of an all-juvenile group of rhesus monkeys. *Amer. J. Physical Anthropol.* 40:83–96.

MacKINNON, J. 1974. The behavior and ecology of wild orangutans *(Pongo pygmaeus).* *Animal Behaviour* 22:3–74.

MERRICK, N. J. 1977. Social grooming and play behavior of a captive group of chimpanzees. *Primates* 18:215–224.

MICHAEL, R. P., and HERBERT, J. 1963. Menstrual cycle influences grooming behavior and sexual activity in the rhesus monkey. *Science* 140:500–501.

MICHAEL, R. P.; HERBERT, J.; and WELEGALLE, J. 1966. Ovarian hormones and grooming behaviour in the rhesus monkey *(Macaca mulatta)* under laboratory conditions. *J. Endocrinology* 36:263–279.

MILLER, M. H. 1976. Behavioral effects of amphetamine in a group of rhesus monkeys with lesions of dorsolateral frontal cortex. *Psychopharmacology* 47:71–74.

MILLER, M. H., KLING, A., and DICKS, D. 1973. Familial interaction of male rhesus monkeys in a semi-free-ranging troop. *Amer. J. Physical Anthropol.* 38:605–11.

MITCHELL, G., and BRANDT, E. M. 1972. Paternal behavior in primates. In *Primate socialization,* ed. F. E. Poirier, pp. 173–206. New York: Random House.

MITCHELL, G., and TOKUNAGA, D. H. 1976. Sex-differences in nonhuman primate grooming. *Behav. Process.* 1:335–345.

MORI, A. 1975. Signals found in the grooming interactions of wild Japanese monkeys of the Koshima troop. *Primates* 16:107–140.

MORRIS, K., and GOODALL, J. 1977. Competition for meat between chimpanzees and baboons of the Gombe National Park. *Folia Primatol.* 28:110–121.

MORUZZI, G. 1969. Sleep and instinctive behavior. *Archives Italienne de Biologie* 107:175–216.

NAPIER, JOHN 1962. The evolution of the hand. *Scientific American,* December.

NEVILLE, M. K. 1972. Social relations within troops of red howler monkeys *(Alouatta seniculus).* *Folia Primatol.* 18:47–77.

NOLTE, ANGELA 1955. Friedland Beobachtungen über das Verhalten von *Macaca radiata* in Sudindein. *Z. Tierpsychol.* 2.

OKI, JOHEI, and MAEDA, YOSHIAKI 1973. Grooming as a regulator of behaviour in Japanese macaques. In *Behavioral regulators of behavior in primates,* ed. C. R. Carpenter, pp. 149–163. Lewisburg, Pa.: Bucknell U. Press.

OPPENHEIMER, J. R., and OPPENHEIMER, E. C. 1973. Preliminary observations of *Cebus nigrivittatus* (Primates: Cebidae) on the Venezuelan Llanos. *Folia Primatol.* 19:409–436.

POIRIER, F. E. 1968. The Nilgiri langur *(Presbytis johnii)* mother-infant dyad. *Primates* 9:45–68.

PRESTON, D. G.; BAKER, R. P.; and SEAY, B. 1970. Mother-infant separation in the patas monkey. *Develop. Psychol.* 3:298–306.

RAHAMAN, H., and PARTHASARATHY, M. D. 1969. Studies on the social behavior of bonnet monkeys. *Primates* 10:149–162.

RANSOM, T. W., and RANSOM, B. S. 1971. Adult male-infant relations among baboons *(Papio anubis). Folia Primatol.* 16:179–195.

REDICAN, WILLIAM K. 1975. Facial expressions in nonhuman primates. In *Primate behavior: developments in field and laboratory research,* no. 4, ed. L. A. Rosenblum, pp. 103–194. New York and London: Academic Press.

RICHARD, ALISON F., and HEIMBUCH, RAYMOND 1975. An analysis of the social behavior of three groups of *Propithecus verreauxi.* In *Lemur Biology,* ed. Ian Tattersall and R. W. Sussman, pp. 313–333. New York and London: Plenum.

RIJKSEN, H. D. 1978. A field study on Sumatran orangutans *(Pongo pygmaeus abelii* Lesson 1827): ecology, behavior and conservation. Wageningen: Veenman and Zonen.

ROSENBLUM, L. A. 1971. Kinship interaction patterns in pigtail and bonnet monkeys. In *Proc. 3rd Internat'l. Cong. Primatology,* ed. H. Kummer, vol. 3, Basel: Karger.

ROSENBLUM, L. A.; KAUFMAN, I. C.; and STYNES, A. J. 1966. Some characteristics of adult social and autogrooming patterns in two species of macaque. *Folia Primatol.* 4:438–451.

1969. Interspecific variations in the effects of hunger on diurnally varying behavior elements in macaques. *Brain, Behavior and Evolution* 2:119–131.

ROTHE, H. 1971. Some remarks on the spontaneous use of the hand in the common marmoset *(Callithrix jacchus).* In *Proc. 3rd Internat'l. Cong. Primatology,* Zurich 1970, ed. H. Kummer, vol. 3, pp. 136–141. Basel: Karger.

ROWELL, T. E. 1966. Hierarchy in the organization of a captive baboon group. *Animal Behavior* 14:430–443.

1968. Grooming by adult baboons in relation to reproductive cycles. *Animal Behaviour* 16:585–588.

1971. Organisation of caged groups of *Cercopithecus* monkeys. *Animal Behaviour* 19:625–645.

ROWELL, T. E.; DIN, N. A.; and OMAR, A. 1968. The social development of baboons in their first three months. *J. Zoology* 155:461–483.

SAAYMAN, G. S. 1971. Grooming behaviour in a troop of free-ranging chacma baboons *(Papio ursinus). Folia Primatol.* 16:161–178.

SADE, D. S. 1972. Sociometrics of *Macaca mulatta.* I. Linkages and cliques in grooming matrices. *Folia Primatol.* 18:196–223.

SAHLINS, MARSHALL D. 1965. On the sociology of primitive exchange. *The relevance of models for social anthropology.* ASA Monographs, vol. 1, pp. 139–236. London: Tavistock; New York: Praeger.

SCHALLER, GEORGE B. 1963. *The mountain gorilla: ecology and behavior.* Chicago: U. of Chicago Press.

SCHEFLEN, ALBERT E. 1973. *Communicational structure: analysis of a psychotherapy transaction.* Bloomington and London: Indiana U. Press.

SEYFARTH, ROBERT M. 1976. Social relationships among adult female baboons. *Animal Behaviour* 24:917–938.

SIMONDS, PAUL E. 1965. The bonnet macaque in south India. In *Primate behavior: field studies of monkeys and apes*, ed. I. Devore, pp. 175–196. New York: Holt, Rinehart and Winston.

1974. *The social primates.* New York: Harper and Row.

SIMPSON, M. J. A. 1973. The social grooming of male chimpanzees. In *Comparative ecology and behaviour of primates*, ed. R. P. Michael and J. H. Crook, pp. 411–505. New York and London: Academic Press.

SOUTHWICK, CHARLES H. 1967. An experimental study of intragroup agonistic behavior in rhesus monkeys *(Macaca mulatta). Behaviour* 28:182–209.

SPARKS, JOHN 1967. Allogrooming in primates: a review. In *Primate ethology,* ed. D. Morris, pp. 148–175. Chicago: Aldine.

STRUHSAKER, THOMAS T. 1967. Behavior of vervet monkeys *(Cercopithecus aethiops). U. of California Publications in Zoology* 82.

1971. Social behavior of mother and infant vervet monkeys (*Cercopithecus aethiops*). *Animal Behaviour* 19:233–250.

1975. *The red colobus monkey.* Chicago: U. of Chicago Press.

STRUHSAKER, T. T., and GARTLAN, J. S. 1970. Observations on the behaviour and ecology of the patas monkey *(Erythrocebus patas)* in the Waza Reserve, Cameroon. *J. Zoology* 161:49–63.

STRUM, S. C. 1975. Primate predation: interim report on the development of a tradition in a troop of olive baboons. *Science* 187:755–757.

SUGIYAMA, Y. 1969. Social behavior of chimpanzees in the Budongo Forest, Uganda. *Primates* 10:197–225.

1971. Characteristics of the social life of bonnet macaques. *Primates* 12:247–266.

SZALAY, F. S., and SELIGSOHN, D. 1977. Why did the strepsirhine tooth comb evolve? *Folia Primatol.* 27:75–82.

TELEKI, GEZA 1973. *The predatory behavior of wild chimpanzees.* Lewisburg, Pa.: Bucknell U. Press.

1975. Primate subsistence patterns: collector-predators and gatherer-hunters. *J. Human Evolution* 4:125–184.

TRIVERS, R. L.1971. The evolution of reciprocal altruism. *Quart. Rev. Biol.* 46:35–57.

VESSEY, S. H. 1971. Free-ranging rhesus monkeys: behavioral effects of removal, separation, and reintroduction of group members. *Behaviour* 40:216–227.

VICK, L. G., and CONLEY, J. M. 1976. An ethogram for *Lemur fulvus*. *Primates* 17:125–144.

WIESENFELD, SUZSANNA; HALPERN, BRUCE P.; and TAPPER, DANIEL N. 1977. Licking behavior: evidence of hypoglossal oscillator. *Science* 196:1122–1124.

YOUNG, MICHAEL W. forthcoming. On refusing gifts: aspects of ceremonial exchange in Kaluana. In *Metaphors of interpretation: essays in honour of W.E.H. Stanner*, ed. D. Barwick et al.

THE LIMITS OF TRANSMISSION

THE EMERGENCE OF MEANING

Culture has usually been regarded as a form of transmission, and the propagation of material objects from one generation to the next makes this process seem self evident. However, even in the case of tools, there is no reason to suppose that cultural content can be transmitted in a literal way. If the mechanism underlying the instrumental modality of action is the ability to abstract from experience the patterns of cause and effect, then the tool is re-created with each new user and the material object itself is simply a vehicle for a process of discovery. By this analysis, observational learning is not information transfer but an abstraction of patterns from experience. Premack's chimpanzee, Sarah, provides a good example of this process (Premack and Woodruff forthcoming).[1]

Sarah was shown four short videotapes of a human confronting a particular situation: (1) a person attempting to escape from a locked cage, (2) a person pretending to shiver in the presence of a heater, (3) a person trying to turn on an unplugged phonograph, and (4) a person trying to wash the floor with the hose not attached to the faucet. At the end of each tape, Sarah could choose correctly among different photographs, one of which was the correct solution, such as photos of an electric cord that was plugged in and intact, a cord plugged in but cut, or a cord that was not plugged in. It is significant that Sarah had never performed any of these actions herself but had nonetheless encoded the actions and effects mentally and could call upon this mental representation to control the selection of the proper object for solution. Since each act is itself a complex sequence, the chimpanzee also had to segment the action series herself as well as encode it. This ability indicates that a chimpanzee can not only learn the cause-and-effect relationships of actions and learn to perform sequences of actions instrumental to a common goal but can *infer* cause and effect relationships by watching others and use the inferred relationships to guide its own action. Thus the basic processes for observationally based coordination of action are present in contemporary apes. Expansion of parietal cortex could be theorized to provide an additional level of nested self-concepts in which the observer was represented to himself in the hypothetical eyes of the other. This postulated nesting is in accord with Gallup's findings on the evolutionary progression of mirror image concepts in primates, and it also explains further peculiarities of the human species. From Goffman's analysis of human social interactions and the human propensity for clothing and self-decoration, unique among primates, it is necessary to infer that the instrumental presentation of self is a normal aspect of modern human behavior. Within the theoretical framework used here, the simulated perception of how others will regard the action must be able to function as a goal of an instrumentality (Goffman 1959, 1961, 1975).

Observational learning, viewed as a process of discovery about another's actions, has quite different implications from a transmission model of culture. In a transmission model, the information handed down has an existence independent of the minds of its bearers, and hence there is no systematic relationship between cultural content and individual psychology. In the traditional transmission model, the observing animals are simply vessels to be filled. If viewed as a discovery process, however, observational learning would be most economically construed as the inferential creation of new conceptual linkages between actions and object concepts already in repertory, based upon perceived novel relationships among exemplars of those concepts. This approach predicts a systematic relationship between the kind of information that can be learned through observation and the information that has already been conceptually stored. It is in language acquisition studies that effects analogous to these are easiest to demonstrate. A child perceives the "same" input information quite differently depending upon its prior linguistic knowledge, and certain distinctions, obvious to an adult speaker, are simply not perceived until prior concepts have been incorporated. For example, in Huttenlocher's experiments with subject position in English, young children would move toy trucks as if they perceived the agent of action as the first noun group in a sentence, even when the word order was reversed with passive transformations (Huttenlocher et al. 1968). That is, the sentences "The green truck is pushing the red truck" and "The green truck is being pushed by the red truck" produced the same action sequence. In the nonverbal sphere, Greenfield and her colleagues found that children of different ages would imitate the same nonverbal actions differently (Goodson and Greenfield 1975; Greenfield et al. 1972). In a demonstration of nesting cups, in which five cups of graduated size were placed one within the other by the experimenter, children spontaneously imitated the action when given the cups. However, the nesting strategies used by the child differed as a function of age, even though all children had observed the same demonstration.

If observational learning is a form of discovery about relevant patterns, and if symbolization already exists in the conceptualization of apes, then it ought to raise questions about culture as a level of organization based upon processes of cultural transmission. An ethological approach would argue that culture is not a transmission process but an interaction process. Moreover, it is not mediated in humans by a capacity for culture at all but by a delicate responsiveness to the needs and intentions of others and extremely sophisticated instrumental and affective mechanisms for asserting needs and intentions in a social context.[2] In this respect too, there is continuity between humans and apes. Savage-Rumbaugh has shown that the copulatory behavior of pygmy chimpanzees *(Pan paniscus)* is normally coordinated through the use of gestures that require one animal to infer the intention of the signaler from the act itself (Savage-Rumbaugh et al. 1977). For example, a pushing gesture on the shoulder is interpreted as a signal for the recipient to turn around. From the signaler's point of view, this gesture is a part behavior derived from the instrumental act of physically turning the other. However, as a social signal, it requires that the recipient infer the intended goal from the morphology of the behavior and then instrumentally implement that goal himself. A more abstract example is touch-plus-iconic-hand-motion gestures, in which the part of the body to be moved is lightly touched and the direction indicated by hand motion in the desired direction. Savage-Rumbaugh has argued that the iconic gestures are derived from the intention movements of positioning, but there is another interpretation that makes the sign language abilities of apes far less mysterious. Theoretically, the iconic type of signal can be viewed as the iconic driving of the hand by an internal image, as already discussed in ape sign language. In the class of completely iconic hand gestures, the indicative touch is missing, and the signal is simply a directional sweep of the hand. Since a sweep of the hand without a chimp in the grip is *not* part of the instrumentality of physically repositioning another animal, it must be programmed in some other way; and the simplest explanation is what Freud suggested long ago: the iconic driving of an effector through the wishful simulation of a desired goal. This interpretation seems justified because, ethologically speaking, these gestures do not quite fit the definition of an "intention movement." An intention movement is a part behavior derived from the initial act of an instrumental sequence, like preparing to get up from the table. The initial act of repositioning is touching or gripping the other. These iconic copulatory gestures, in contrast, are closer to *anticipatory movements,* which are responses to an event that has not yet happened. The ladder building of chimps filmed by Menzel provides clear instances of moving in anticipation of the consequences of an action, such as making the preparatory movements of climbing a ladder before the ladder is in place (1972; Reynolds forthcoming). From the perspective of motor behavior, they are a kind of mistake, for clearly a sweep of the hand will not move a chimpanzee, nor can an unpositioned ladder be climbed. Yet in a social context, the behavior appropriate to a completed action can serve to cue that action in an intelligent perceiver. Psychologically, therefore, the touch plus iconic gestures appear to be similar to the begging gestures of chimpanzees, often described, which are derived from the act of receiving rather than from the act of taking. More precisely, the begging gesture is derived from the act of anticipating receiving, because no food has yet been offered. Behavioral anticipations, therefore, require simulation of the consequences of action, in either the run-off of behavioral programs in contexts to which they are inappropriate or run-off in the absence of fulfilled "entry conditions," which normally constrain the onset of an act. In any case, there is a kind of loosening of the reality constraints that normally operate on adult behavior, and this is in accord with the commonly accepted definition of wishful thinking. Although this can be a pathological disruption of action from the point of view of integrated motor performance, it is adaptive if there are other individuals who respond to wishes as if they were intentions. The decoding of such signals requires an animal capable of perceiving context-free patterns (a sweep of the hand) and relating them to context-sensitive patterns (the location of the other individual relative to oneself) and to inferential information (what his goal was). This is indicated by the fact that the same hand sweep can mean different things depending upon the location of the signaler —for example, either to approach or to go past. Consequently, the coordination of copulatory movements in pygmy chimpanzees exhibits the kind of goal-directed intentionality characteristic of human nonverbal behavior, and it indicates that instrumentality and affect are a coordinated action system in this species as well.

Since chimpanzees can apparently only drive body parts with an internally generated image, and not the constructional process itself, the ape signaling system is limited to movements of the body or to extensions of the body by means of the gripping action. Humans, however, are also capable of producing constructions that simulate the form of internal images. It is this advanced sensory-motor system, working in conjunction with innate images of the affective modality, conceptually represented, that produces those striking morphological convergences between the images of animal display and cultural product pointed out by ethologists. The output program is specifically human, but the resulting product is indeed phylogenetically homologous with innate perceptual mechanisms. Contrary to an instinct theory, however, the output program is volitional motor, as the cultural variability in the distribution of such innately modeled images indicates; and their effect is not to release behavior but to operate on the affective component of the conceptual system. Their propagation within a society no doubt results from the common perception of their causal efficacy in dealing with others in certain affective contexts. It is this capacity to incorporate the constructional aspects of instrumental action into social situations that also provides a partial mechanism for the phenomenon that Erving Goffman has emphasized time and again — the fact that humans do not simply interact but actively construct the framework within which interaction takes place. Even more important, from a phylogenetic perspective, the chimpanzees provide the first glimmering of communication based on prestational exchange in which wish and instrumentality reinforce each other.

If the information content of culture is not transmitted but discovered, through the interaction of external information and internal process, then it is clear that "traditional transmission" is not a property that can be predicated of either learned information or of the mechanisms of information acquisition. It is an attribution in a metalanguage that presupposes knowledge of both the action produced by the model and the subsequent action produced by the imitator. "Traditional transmission" is a relational concept in a metalanguage of action exemplified by a congruity in the actions of the model and the imitator engaged in observational learning. To suppose that this concept denotes distinct mechanisms at the level of psychological process is to make exactly the same mistake with culture that McDougall made with instinct: because a mechanism performs a certain function, it does not mean that the function itself is encoded. Just as reproduction of the species is an effect of sexual mechanisms that need have no knowledge of their own significance, so traditional transmission is an *effect* of observational learning and not a mechanism in itself. The major question for human evolutionary theory, however, is whether these conclusions are still justified as a genuinely linguistic organism. In the following section, it is argued that language too is an interaction process whose major function is not the transmission of information but social coordination implemented through the induction of belief in the minds of listeners.

THE MAGIC OF WORDS

As is well known, the era of cognitive psychology received its impetus from the linguistic critique of Skinner's theory of language acquisition, and primatological studies have since paid obeisance to the uniqueness of language. However, the uniqueness has often been characterized in terms of Hockett's now famous list of the "design-features" that distinguished language from other forms of primate communication: the *arbitrariness* of the relation between sign and referent, the ability to talk about things removed in space and time *(displacement)*, the formulation of statements on which truth or falsity can be predicated *(propositionality)*, the emergence of a level of meaning *(semanticity)*, the encoding of new information and new propositions *(productivity)*, and the encoding of information according to certain rules *(grammaticality)*, to take the most general ones (Hockett 1960; Hockett and Ascher 1964). However, if we examine the evidence of primate cognition already available, particularly in the language analog experiments, there is far more continuity than an emergent capacity for symbolization would lead us to expect. Concepts, after all, allow the encoding of information removed in space and time; they allow the formulation of hypothetical constructs amenable to the predication of truth and falsity; they are productively extended to new situations; they are lawfully interrelated; they are combined one with another; and they stand in relation to the perceptual world in exactly the same way as a sign stands to its referent.[3] There is already more correspondence than can be explained away as convergence, and Premack himself suggests that the most important design-features of language are not the properties of language at all but of the underlying conceptual processing, homologous in ape and man. What is left unexplained is the *arbitrariness* of sign and referent — that Maginot line of linguistic philosophy.

Premack and Rumbaugh specifically designed their experiments to incorporate sequential ordering and arbitrary equivalences because it had been so often claimed that any convincing language analog would have to have these properties. The chimps have in fact learned to deal with arbitrary equivalences between sign and referent, but the critic could easily argue that these arbitrary symbols were not arbitrary at all since they were acquired in a response-contingent way. The tags may have been arbitrary to the experimenter, but they were conveyed to the animal in a way that was not arbitrary but causal, since particular combinations invariably produced the same results. Nonetheless, if we go beyond the chimps' performance to the implicit philosophy that gave rise to these experiments in the first place, the result is still profoundly unsettling to the claims for a specifically human faculty of symbolization: if the critic's argument can be applied to chimpanzees, it can equally well be applied to human beings. There is not only the growing awareness of the extent to which humans are locked into cycles of mutual response contingency, best conveyed by studies of mother-infant interaction, but not much reflection is needed to conclude that a language in which the relation between sign and referent were arbitrary would be unlearnable and unintelligible. Once again, the "arbitrariness" of sign and referent is a predication in a metalanguage in which the psychological contingencies of actual speech are conveniently ignored. Because *dog* is conveyed by *Hund* in Germany and *chien* in France, it does not follow that the relationship between the concept dog and its speech encoding is arbitrary to speakers of German and French. To the contrary, the relationship between word and concept is always response-contingent in any population of speakers — and could not be transmitted if it were not. Arbitrariness is not a psychological concept at all, but a linguistic one, purchased at the very high cost of excluding the actual processes by which language is learned and used. What Hockett called arbitrariness is a poor name for the obvious fact that the response-contingent relationships between concepts and speech sounds differ among different human populations. The relevant design-feature is the *response-contingent linkage* between vocal act and conceptual inference, and this reformulation is a more radical departure than might at first be apparent. Psychologically speaking, it implies that the chimpanzee's inability to grasp the relationship between sign and referent without caus-ally efficacious contingencies cannot be used as proof for a symbolic faculty in man and its absence in the chimpanzee. On the contrary, the most par-simonious conclusion is that humans could not learn to use tags either if their use did not have causally efficacious results. In the reluctance to grant referential competence to chimpanzees, we encounter once again the long shadow of nineteenth century scientism with its belief that causality has some necessary connection to physics. In rejecting the arbitrariness of the relationship between sign and symbol, one does not therefore have to fall back on physicalist theory of reference. In the realm of communication between animals, it is quite possible to infer causal relationships between signal and response that owe their constancy not to the intrinsic properties of the signal but the repeatability of the reaction it provokes. Consequently, in asserting the response-contingent nature of linguistic acts and rejecting a specifically human symbolic capacity, one is not expanding the domain of physics but embedding language firmly in the substrate of primate cogni-tion.

If referential communication in man is the outgrowth of conceptu-ally generated acts of which the consequences on others are in turn assessed by the conceptual mechanisms of cause and effect, then the mechanisms of artificial language in apes are partly homologous with the mechanisms of language in humans. What is experimentally contrived is the perceptual situation and not the cognitive abilities that the ape uses to interpret it. If a tag is regarded as a member of a disjunctive equivalence class, a class con-taining a tag and its associated concept, then psychologically speaking there is no reason to regard tags as any more (or less) complicated than disjunctive instrumental concepts generally. In other words, when an ape learns that two dissimilar objects, a rock and a beach ball, can both be thrown, it is likely to have formed a disjunctive instrumental concept embracing both rock and ball—a node that links them both to a common instrumentality. If tags are simply instrumental concepts, then the language analog experiments are not artificial at all but completely natural. Teaching an ape an artificial sign is methodologically comparable to the standard physiological technique of substituting a radioactive isotope for one of the atoms in a compound normally used by an animal. The radioactive substitute is processed nor-mally by the organism's metabolism, but its artificiality makes the normal pathways of reaction visible to the experimenter. Symbolization, in other words, does not evolve in hominids: it already existed as the capacity to form disjunctive instrumental concepts, and it is this capacity that the language analog experiments bring to light.

The cognitive approach to language ability, by focusing on the goal-directedness of behavioral productions, has the advantage that it uses the same body of theory to explain both language in humans and sign language in apes, and it also postulates coevolution between language and tool use (Hewes 1973b; Lenneberg 1960; Greenfield and Smith 1976; Bruner 1973; Reynolds 1976; Ratner and Bruner 1978). The theory can therefore explain the detailed correspondences between language and tool use, which several authors have pointed out. Both activities in primates are under volitional control, and the abstract form of tool use can be impaired by cortical lesions to the speech hemisphere in man (Kimura and Archibald 1974). Moreover, the structural properties of tool use match language in a detailed way. Tool-using behavior is hierarchically organized into nested levels of control, which can be easily illustrated by everyday skills. In striking a match, for example, each hand performs a tonic grip with an object, and each configuration is used as a unit at the next level of combination—the relation of the objects to each other. The match, held in one hand, is brought to the striking surface on the matchbox held in the other, and another tonic behavior must hold the two objects in contact while the strike is made. It is not the match that makes the strike but the configuration of "hand-that-is-holding-a-match-and-holding-it-in-contact-with-the-striking-surface." Furthermore, tool use is productive in that different objects can be incorporated into the same program as seen particularly in play but also in circumstances in which a suitable tool is unavailable. Tool use is also productive in the sense that the same unit can participate in different programs, as when the unit "hand holds hammer" is put to work as a lever, a club, or a rake. Tool use also is affected by emotional state, in a manner analogous to paralinguistic phenomena, which affects the speed and organization of the performance. Also the basic categories of skilled action —location, agent, target, and so forth —have a parallel in the grammatical categories of natural language.

The *instrumental theory of language,* as the coevolution of speech and tool use can be termed, would explain these correspondences by the hypothesis that language and tool use employ the same mechanisms of motor output and conceptual processing and by the fact that both are acquired through observational learning in conjunction with response-contingent activity. However, Lenneberg's case of a child who had good verbal understanding in spite of a congenital abnormality of vocal control preventing the articulation of speech sounds suggests that even self-generated response-contingency may have been overrated and that language contingencies are very accessible to observational learning (1962, 1964; Palermo 1975). A combined observational learning and instrumental theory would even predict that apes could acquire a good deal of comprehension of spoken speech even though they are incapable of producing it.[4] Nonetheless, the theory still has shortcomings as a general model of language, which can be demonstrated by simply traveling to a foreign country or viewing a foreign film. The instrumental action sequences observable on the visual channel are generally recognizable and comprehensible to a human being from any culture, whereas the acoustic channel is totally incomprehensible unless one knows the language. The most obvious criticism of the instrumental theory is that the sign language performances of apes are based on these panhuman commonalities of the visual channel and therefore have nothing to do with language processing at all. This criticism of the instrumental theory is given greater force by the fact that when language uses the visual channel, as in writing or natural sign language, these components of the visual channel are every bit as incomprehensible to the untutored observer as is speech itself. Moreover, there is the natural component of speech processing already encountered in the discussion of phonemic perception in man, which also argues for a distinct linguistic system not reducible to general cognition.[5] As cogent as these arguments are, they are not necessarily fatal to a theory of language based on cognition and tool use. What these arguments ignore is the possibility that both language and tool use can use exactly the same psychological mechanisms and still produce very different results because the input information is subject to different constraints. There is a great deal of evidence indicating that the constraints operating on language acquisition could be very different from those operating on tool-use acquisition, even though both categories of action have access to the same cognitive mechanisms.

The most pertinent fact about language is that it is a system of social action: unlike tools, utterances have no external effects whatsoever if they are not acted upon by others. This fact indicates that language should not only have the properties of instrumental action but the properties of social action as well. First of all, language should not be an exclusively instrumental form of action but should require the joint participation of the affective modality — the system specialized for the assessment of social consequences. Not surprisingly, emotional indicators are ubiquitous in speech, and it is generally acknowledged that words not only have a denotative meaning but a connotative meaning as well. Osgood's studies using the semantic differential test, which demonstrate cultural differences in attitude toward words with similar denotative meanings, and the effect of connotation on brain evoked potentials, illustrate this fact quite well (Chapman et al. 1977; Osgood 1964; Osgood et al. 1975). Second, the response-contingencies of language, because they are constrained by the social responses of others and not by the constraints of *physical* cause and effect, are potentially crossculturally variable in a way in which tool-using programs are not. In other words, the selective factors that constrain the variability of goal-directed action differ in language and tool use. Tool use is highly constrained by factors that can be expected to be uniform among different populations of humans, whereas there is no comparable physical constraint reducing the variability of linguistic acts save universal properties of human psychology. Moreover, there are positive factors selecting for systematic variability in language, since utterances that are both novel and interpretable can be expected to have disproportionate attention-getting power. Also, since language is lacking in universal constraint relative to tool use, it is especially suitable for creating social distinctions that reinforce group differences, a point on which both students of learned bird song and sociolinguistics are in agreement (Nottebohm 1970).

The fact that different "selective factors" are operative in the acquisition of programs for language and tool use would lead to different degrees of cultural variability in the two forms of action even if the mechanisms were identical. Tool use would be constrained by the universally applicable constraints of physical causality, whereas language, lacking such selective forces, would constantly diverge from its existing state and be uninterpretable to the outsider. To reject the instrumental theory of language, it is first necessary to show that the differences in selective constraint between tool use and language are still inadequate to account for the differences. According to the instrumental theory, the constraints of language are derived from social response-contingencies and hence are comparable to a social ritualization process. Fortunately, humans also perform other kinds of ritualized behavior, which is nonlinguistic and observable through the visual channel, and this could provide a test case for the cognitive theory. If we observe a ritual in a foreign culture, or even our own, such as a Catholic mass, the sequence of actions may be as incomprehensible as a foreign language. The individual actions are readily identifiable — breaking a wafer of unleavened bread, uttering quotations from Scripture, bowing to the altar, and so forth — but the composite act is still not readily interpretable. This indicates that ritualized behavior can be both crossculturally variable and incomprehensible in a way that is reminiscent of language (Douglas 1972). However, the instrumental theory of language immediately encounters another difficulty, for ritual acts and linguistic acts differ in one significant respect. For linguistic acts, it is always possible to translate the sentences of one language into the sentences of another (Steklis and Harnad 1976). Even accepting the caution that translation always distorts, it is nonetheless a fact that the world is full of bilinguals who manage to make the sentences of one language understood in another. In this respect, language is quite unlike any other form of ritualized social act, for in what way does a Catholic mass have a translation into another ritualized system of acts? Even within the auditory modality, in what sense can a jazz composition be said to be translatable into the music of the Renaissance chapel?

One of Hockett's design-features of language was *interchangeability,* which he defined as the ability to produce what can be understood, and he contrasted it to systems in which sender and receiver used different signals, as in courtship displays. It is clear that any socially ritualized system acquired through observational learning would have this property, and chimpanzees can in fact use their acquired sign systems to communicate with each other (Savage-Rumbaugh et al. forthcoming). However, the feature of *translatability* makes a far stronger claim for linguistic uniqueness because it suggests that something remains invariant through the transformation from one coding to another, namely the meaning of the message. The fact of translation provides good prima facie evidence for a distinct linguistic capacity, but a consistent instrumental position would argue that translatability is another metalanguage concept, based upon abstraction from the behavior of bilingual persons who know not only two different rule systems but also those productions that are socially equivalent on the basis of the situations in which they can be used. Although the experiment has yet to be tried, the instrumental theory of language would predict translatability in chimpanzee sign language: that a chimpanzee could acquire a second code by being immersed in a different sign system and would be able to produce a novel but matching production in one code when given a novel example in another. If a chimpanzee could perform this task, then the instrumental theory, which postulates equivalence classes relating rule-generated productions, would be considerably strengthened. Assuming apes can do this, the instrumental theory, while successfully disposing of the translatability objection, would make a prior problem even more disconcerting. If both apes and humans are creatures prone to rule-generated acts acquired in the course of participation in and observation of response-contingent social behavior, why should one such system be meaningful when the others—ritual, games, and art—are so psychologically different in spite of such similar properties? The most troublesome fact of all for an instrumental theory is that all of these activities can be as volitional, goal-directed, structurally complex, and socially ritualized as language, but only language appears to have a direct path into the conceptual system of its perceivers. Whereas other ritualized and productive behavioral systems are observed and interpreted by humans, *only linguistic productions* are commonly interpreted as productions for which the predication of truth and falsity is meaningful. In other words, language is a form of behavior with the properties not only of the instrumental modality but the properties of conceptual information itself. It is not surprising it has proved so refractory to phylogenetic explication.

In view of the fact that purely morphological definitions of behavior have been found wanting in the characterization of tool use, it is possible that they are equally inadequate for language. If the fixed-action patterns of monkeys can be manifestations of sophisticated control processes far in advance of their behavioral form, then it is equally possible that the evolutionary precursors of language are not be found in the morphologically similar behaviors of the instrumental modality but in a system that provides a direct pathway to conceptual reorganization. In other words, since language uses the output and interpretive mechanisms of the instrumental modality, the instrumental theory is correct in asserting a systematic relationship between language and tool use, but since distinctively different systems can channel information in this way, the instrumental theory is incorrect in reducing language to the properties of the instrumental modality alone. Language could still be a biologically distinct system even if all of its *behavioral* attributes were explicable in terms of the input–output mechanisms employed. Conversely, behavior might still be linguistic even if it shared none of the productive and sequentially complex features associated with the primate instrumental modality, as in swearwords and other overlearned linguistic formulae. Language could be a distinct central system which usually employs the advanced sensory and motor capacities of the primate brain but is not reducible to them. The *instrumental channel theory* would argue that language must be defined in terms of a specific central system which could be manifested differently at different phyletic levels.

If language is a distinct system, it ought to be localizable in some distinct neural mechanism, and the simplest hypothesis is that it is closely related to a specific sensory modality. The peculiar properties of phonemic hearing argue for a specific speech decoder, as Liberman has maintained, and the historical conservatism of phonemic systems, the presence of laterality, and the phonemic powers of neonates all support this interpretation (1970; Liberman et al. 1967). However, an argument used against this view is the apparently arbitrary relationship between language and sensory modality. Three lines of evidence have been advanced in favor of this proposition: (1) that language can develop in a nonauditory channel among the deaf; (2) that apes can learn linguistic analogs in the visual–gestural channel; and (3) that language can be extended to other channels in speaking humans, as in writing and in Braille. These observations have been further used as evidence for the hypothesis that language in fact has switched sensory modalities in the course of hominid evolution (Hewes 1973a; Kendon 1975). Since the visual channel is so much better developed in apes, it has been argued that language was first visual–gestural and only later became vocalauditory. This hypothesis multiplies the phylogenetic difficulties. It not only does not explain why modern apes are not linguistic when they are postulated to have the necessary skills, but it requires a further channel switch, which is also left unexplained. Since apes do not normally have language and humans normally have it in the auditory channel, phylogenetic speculations should derive the hypothetical past from the existing present and not the other way around. Even by the theory's own logic, if language is in fact a distinct central system that can use the instrumental modality as an input–output device, focusing on the permutations of sensory modality is beside the point. However, if one is interested in accounting for the phylogeny of the central system itself, then any preferential connection between language and a particular sensory modality cannot be dismissed as a historical curiosity but is itself *the central historical fact*. Language is primarily vocal-auditory in the one species that has it, and it does not occur in related species that do not have learned vocal-auditory behavior, however sophisticated their cognition might otherwise be. This suggests that language is a phylogenetic derivative of the primate auditory system and its central connections. By this reasoning, language in contemporary humans would be postulated to engage a phylogenetic derivative of the auditory system regardless of the sensory modality of the input information. In an auditory-derivative definition of language, language would always be "central speech," even in the congenitally deaf.

The simplest version of this auditory channel hypothesis would locate language in the auditory association cortex, but this formulation is too simple because language is known to have an extensive cortical representation that exceeds the boundaries of the auditory system. Electrostimulation of the exposed brains of conscious human patients with a history of brain disorder produces interference with speech from a wide range of locations in the left hemisphere (Penfield and Roberts 1959). Arrest of ongoing speech, distortion and repetition of words, hesitation and slurring, inability to name, and counting errors are produced from a wide band across the left hemisphere, from the posterior portion of the third gyrus of the frontal lobe to the posterior portion of the parietal lobe. However, the responses concentrate in the cortex immediately adjacent to the mouth region of the precentral cortex of the frontal lobe, traditionally called Broca's area, and around the posterior part of the sylvian fissure, in temporal-parietal cortex, traditionally called Wernicke's area. The evidence derived from the study of brain lesions in man has produced a similar picture. Disorders of speech and language are likely to be produced within this general area, with the permanence or inevitability of aphasic impairment questioned for Broca's area but generally conceded for the posterior temporal-parietal cortex (Bogen and Bogen 1976; Luria 1964; Pribram 1971; Zangwill 1975). Because these language areas of cortex receive their input from different thalamic nuclei lying subcortically and are interconnected by transcortical "association" pathways, it is not surprising that lesions in different portions of the left hemisphere affect different aspects of language function. Lesions in the auditory association cortex of the temporal lobe, area 22, almost always produce severe impairment of phonemic hearing, whereas lesions further back into the angular gyrus of the parietal lobe are more likely to affect visual aspects of language. Clinically, it is possible to characterize different kinds of aphasia depending upon the particular configuration of abnormalities. Roman Jakobson, in evaluating the clinical data from a linguistic perspective, argued that Luria's aphasic types were compatible with a composite process model that postulated operations for both the selection of linguistic units and their combination in both encoding and decoding mechanisms (Jakobson 1964; Jakobson and Halle 1956). The language area apparently includes functionally distinct cortical areas, which produce disassociable impairments on different aspects of language function. This suggests that language involves the integration of largely preexisting cortical systems. Although this makes evolutionary sense, it leaves unanswered the question of why language should be a psychologically distinct behavioral system if it simply involves the multimodal integration of distinct sensory and motor systems. That, it would seem, is precisely what apes are capable of doing.

When Geschwind first argued that humans had language because they could make "intracortical associations," defined operationally as cross-modal transfer, it was believed that nonhuman primates lacked this ability (1964, 1965, 1970).[6] Language may presuppose it, as Geschwind had originally argued, but if language were simply intracortical integration, apes would now be expected to have it too. Geschwind, however, had also argued that the neocortex in man and ape is not exactly comparable because in man there is expansion of the athalamic association cortex of the lower parietal lobe in the vicinity of the supramarginal and angular gyri. Geschwind regarded this region as an association cortex of association cortices and concluded that "in a sense the parietal association area frees man to some extent from the limbic system" (1965, pt. 1, p. 274). However, in the very next sentence he grudgingly admitted: "This independence is only relative since ultimately learning still depends, even in man, on intact connections with limbic structures."

Following the dictates of the Victorian brain, the linguistic system has always been regarded as localized in the most advanced neocortical structures, and the ultimate strategy of the neurology of language has been to locate a region of neocortex in man that is lacking in apes. However, the contemporary neuropsychological evidence supports a very different conception of the brain. The concept of a cortical lesion is itself largely a neurological convention because destruction of neocortical tissue destroys cell bodies in the thalamic areas that project to it, and it has been shown that thalamic nuclei interact with each other (Frigyesi et al. 1972, p. 200 passim). Retrograde degeneration of the thalamus, in fact, is a standard anatomical technique for tracing thalamic projections to the cortex. The Victorian neurologists knew very well that emotional and overlearned phrases could be preserved in the face of severe aphasia, but since these utterances were not propositional, they were not considered relevant to the question of language localization. More pertinent, Penfield and Roberts reported an aphasic patient with a thalamic lesion, and they stressed the close anatomical relationship between the language areas of the neocortex and the underlying thalamus, particularly the pulvinar nucleus (Penfield and Roberts 1959, p. 215). Geschwind, however, argued that the thalamic lesion in question could be producing its effects by fluid pressure outside the thalamus itself.[7]

However, other reports of thalamic involvement in speech were published in the 1960s, after a spate of surgically induced lesions to alleviate neurological disorders (Riklan and Cooper 1977; Ojemann and Ward 1971; Kornhuber 1977; Berlin 1977; Bell 1968). Tests of verbal fluency have since shown laterality effects after thalamic lesions in the ventrolateral nucleus, which sends cerebellar and basal ganglia information to the cortex, and in the pulvinar, the projections of which overlap Wernicke's area by almost everyone's definition of the latter. Since asymmetries have been reported for the neocortex, it is not too surprising that functional asymmetries should occur in the thalamic projections of those areas. Berlin, carrying the causality of speech asymmetries down a step further, has postulated an interaction effect in the thalamic nucleus which projects to the primary auditory cortex, the medial geniculate body, to explain the functional asymmetry between the two ears revealed by the simultaneous input of verbal and nonverbal material. The students of brain lesions in monkeys have also emphasized the role of the cortico-subcortical connections in understanding the function of primate association cortex (Pribram 1972; Myers 1967). Bilateral lesions of the cortex of the inferior temporal lobe abolish learned visual discriminations in monkeys, but this ability is preserved even if the inferotemporal cortex is surgically crosshatched to destroy intracortical association fibers. Undercutting the inferotemporal cortex, however, is equivalent to removing the inferotemporal cortex itself. This association cortex receives projections from the pulvinar nucleus, the projections of which overlap Wernicke's area in man, but the discrimination deficit is not reproduced by pulvinar lesions, indicating that a direct inferotemporal-pulvinar loop is not the immediate source of the difficulty. The inferotemporal cortex also projects to basal ganglia, classically considered to be structures in the extrapyramidal motor system, and basal ganglia influence on the sensory nuclei of the thalamus has been demonstrated electrophysiologically. These various lines of investigation, in both monkeys and men, are pointing toward a very different conception of the neocortex than that inherited from Victorian neurology. Where the Victorians saw an association cortex control system superimposed upon the primitive extrapyramidal (striatal) motor system on one hand and the primitive emotional centers on the other, late twentieth century accounts are seeing loops within loops: cortico-thalamocortical circuits are established, and since cortico-striatothalamic connections are known, the way is also open for bringing this connection full circle. This interactionist conception of higher control functions has also had a significant effect on the postulated relationship between the neocortex and the limbic system structures involved in visceroautonomic and affective phenomena, and by implication, it heralds a changing formulation of the nature of language.

The intracortical association arc emphasized the matching of tag and concept as the substrate of linguistic ability. This emphasis fundamentally misconstrued the naming process, which is functional only to the extent that names can enter into new conceptual relationships (Bronowski and Bellugi 1970). The psychologically significant associations are not those between a tag and its concept but those interlinkages by which conceptual relationships are created and transformed. In his monograph on aphasia, Freud's own concept of a *word* is already a long way from the sign and referent of so much semantic discussion (1953). In the light of what is now known about the encoded nature of phonemes, with their contextually sensitive cues and rules of combination, Freud's word is still a gross oversimplification. Any psychologically adequate concept of the word must contain additional links that connect the perceptual images to programs of encoding and decoding and links to experiential associations, both mnemonic and evaluative. From the perspective of primate psychology, a word is always a social hook on which hominids hang experience.

The word as mnemonic network, most explicitly pursued by Freud in his psychoanalytic work, postulates an inseparable relationship between language and memory (1975). It is this connection, emphasized by psychology but largely ignored by linguistics, that provides a partial solution to the morphological definition of language. Bellugi and Klima, confronted with the problem of evaluating the linguistic nature of American Sign Language among the deaf, used memory intrusion errors as a criterion (1975, 1976; Bellugi et al. 1975). In previous research, with speaking subjects, the nature of the errors in retrieving information from short-term memory had led psychologists to conclude that phonological cues were used in storage. If asked to recall lists of words, for example, the subjects' errors would often be words that sounded like the target word rather than ones that were semantically similar to the target words. Applying this same technique to congenitally deaf ASL speakers, the subjects were given lists of signs presented on videotape with instructions to either write down the English alphabetic equivalent or to repeat in signs. A group of English speakers were given the vocal version of the same test. The memory errors of the two groups were completely different, but they were systematic within each group. The English speakers substituted words that sounded similar, whereas the ASL speakers substituted words that were gesturally similar in their systemic-formational aspects. Neither group recalled extraneous words that were only semantically related to the target. This is especially significant in the case of signers, because there is the further possibility of iconic association—signs that happen to resemble a nonsign configuration of the hands, analogous to "sound symbolism" in speech. These experiments demonstrate very clearly that ASL is a language because the systemic-formative features at the lowest level of combination—the gestural analog to the phoneme (or *chereme*) — are related to short-term memory.[8] Language, in other words, is not only phonemic, whether with the ear or eye, but its phonemic units are also the functional units of a short-term memory device.

On the basis of psycholinguistic study, the memory mechanism can be more precisely specified. As the existence of alphabetic writing systems proves, the phoneme can be considered to be a unit at one level of analysis; but both motor and perceptual studies have shown that a phoneme is a relationship between a number of distinctive features characterized in terms of their articulatory processes: a closed or open velum, the position of the tongue, the vibration of the vocal cords, the obstruction of the vocal passages, and a number of others. In signed speech, similar parameters of variation also occur, and both the articulatory and gestural systems of language can be encompassed by Bellugi and Klima's term of *systemic-formative* features.[9] It is worth reemphasizing that the features are defined in *motor* or articulatory terms, and perception studies have shown that the phonemes are *not* unitary gestalts of sound, because the same phoneme has different acoustic cues in different verbal contexts. Since the intrusion errors of short-term memory tests consist of phonemic substitutions, like *dot* for *tot,* rather than semantic substitutions, like *child* for *tot,* human language can be thought of as a system that has as one of its properties the use of systemic-formative code as a short-term memory store. Furthermore, this systemic-formative code, while sharing features among all existing languages, is itself organized in a culturally variable way. In spite of all the linguistic productivity devoted to the theory of language, there is still no adequate explanation of exactly how a systemic-formative code can be used as a node in an information store or how culturally variable systemic-formative codes can develop in the absence of the overt motor activity they are supposed to produce. However, the presence of intrusion errors based on feature substitutions in the gestural productions of apes would certainly support the genuine linguistic nature of their performances (Thompson and Herman 1977; Hamilton 1974). It would show that the word was not a gestural gestalt but was generated through a codified feature system, and it would also show that the systemic-formative code was systematically integrated with memory.

Somewhat paradoxically, the absence of intrusion errors in ape sign language would not necessarily mean that they lacked a capacity for relating systemic-formative codes to short-term memory. It may only mean that systemic-formative codes bear some special relationship to audition, and that this relationship only becomes manifest when the auditory system is integrated with motor mechanisms. Temporal and sequencing information have been claimed to be especially pertinent to audition, and it is known that unilateral lesions of auditory association cortex in macaque monkeys disrupt auditory sequence discriminations, even though bilateral lesions are usually needed in monkeys to produce discrimination impairments in the cortical areas subserving vision.[10] The best evidence for functional laterality in monkeys also involves auditory sequence discriminations. Warren's postulation of a distinct mechanism for the perception of sound sequences in man, as distinct from sequence discrimination based upon changes in the sound gestalt as a consequence of reordering, which monkeys can certainly do, does not contradict the preadaptation of the auditory channel but extends it (Warren 1976; Waters and Wilson 1976). This evidence raises the possibility that the auditory modality is preadapted for the development of processing mechanisms that require the reorganization of information held in short-term storage on the basis of subsequent information. Since such a hypothetical process bears an intriguing formal similarity to linguistic comprehension, it would be worth pursuing the possibility that the integration of the vocal tract into volitional action was essential to the development of language phylogenetically.

The motor control of the vocal apparatus, volitional and neocorticalized in man, uses cranial nerves as the primary peripheral motor pathway (Hollien 1975). The movements of the jaw are mediated by nerve V, the trigeminal, already encountered in the consummatory behavior of aggression and feeding and influenced by hippocampal stimulation. The movements of the lips and cheeks are controlled by nerve VII, the facial, which already mediates facial expression in primates and hence is responsive to the limbic system generally. Nerve IX is relatively unimportant, since it only supplies the stylopharyngeus muscle. However, nerve X, the vagus, whose visceroautonomic functions are well known, supplies nearly all the laryngeal muscles. Nerve XI, the accessory, innervates portions of the velar musculature and travels in conjunction with the vagus to the muscles of the larynx and pharynx. Nerve XII, the hypoglossal, of which the probable homolog in song birds is lateralized in the control of learned vocalization, controls nearly all of the muscles of the tongue. Speech is not controlled exclusively by cranial nerves, as spinal nerves are involved, primarily in breathing. However, the most important effectors involved in speech are mediated through nerves V, VII, IX, X, XI, and XII. As comparative anatomists have pointed out, all of these cranial nerves form a natural group since their motor components innervate the branchiomeric musculature, derived embryologically from visceral mesoderm, in contrast to the somatic derivation of other skeletal muscle (Webster and Webster 1974, pp. 126, 142ff). This fact in itself may be coincidental, but behaviorally there is no doubt that these same neuromuscular components are already closely integrated with limbic system function in all species of primates. Furthermore, the special sensory systems associated uniquely with the oral cavity are taste and olfaction, and the close anatomical relationship between the limbic system and the olfactory path is well established. These facts suggest that the incorporation of the vocal tract into refined behavior, an event apparently almost entirely restricted to humans among existing primates, can by no means be assumed to be without further consequences neurologically. The development of language in the vocal channel —an event known to have occurred—would require the use of an effector that is already very busy with the control of facial expression and its visceroautonomic correlates. Since this form of communication does not disappear in humans but is made volitional and further differentiated, and since it is required in exactly the same face-to-face settings in which speech occurs, it is clear that two control systems must jointly operate the same effector. We might *expect* cerebral lateralization with speech integrated with one cerebral hemisphere and the affective expressive system with the other (Schwartz et al. 1975).[11] It is important to stress that this dual monitoring does not require that emotion be in one hemisphere and language in the other, for language can be as affective as any other form of behavior. What is required is that the linguistic and facial expression control of shared musculature be functionally specialized, since both forms of behavior must be coordinated simultaneously.

There is also some comparative primate evidence that indicates a behavioral preadaptation of the vocal channel. Green has studied the circumstances under which *coo* sounds are uttered by Japanese macaques (Green 1975a, 1975b). These calls are tonal, and on the basis of sound spectrograms, seven types of coos are empirically distinguishable. These types are correlated with ten distinct classes of social event with some overlap in the context of sounds: (1) an adult male separated from the troop; (2) a female who has lost an infant or a female without an infant during the birth season; (3) a sexually active female temporarily neither consorting nor soliciting; (4) a female approaching or attending to an infant or child; (5) a dominant making affinitive contact with a subordinate; (6) a young animal alone; (7) individuals out of visual contact with group; (8) a calm youngster near or following its mother; (9) calm approach or affinitive contact of a subordinate to a dominant, excluding sexuality and young-to-mother; and (10) female in estrus in a solicitation or consort relationship. These calls are those generally associated with a desire to make affinitive contact, but it is significant that different categories of social contact are to some extent encoded in the call. In fact, many of the basic axes of primate social relationship are represented here: female with or without young, mother and child, adult male, sexually receptive female, subordinate and dominant. Other sound types also occur in the vocal communication of macaques, and Green argued that a class of sounds called *girneys,* produced with an articulatory component of lip and tongue movements, were the homolog of human speech sound articulation. These occur in grooming solicits, as lip smacking, and in approaches by subordinate animals to dominant ones. These calls are often reciprocated, suggesting that they are also a vocal form of prestational exchange. On evolutionary grounds, the coos and the girneys both show preadaptation of the vocal channel for speech. In one case, the categories of basic social relationships appear to be reflected in systematic differences in the morphology of the vocalization; in the other case, vocalizations appear to have been incorporated into the instrumental modality of action where they conform to the dynamics of a prestation process.

Also relevant is Richman's study (1976) of the vocal productions of gelada monkeys *(Theropithecus)*. Old World monkeys and apes, like humans, produce their vocalizations by a glottal sound source, which is further filtered by the configuration of the supralaryngeal vocal tract to produce distinct acoustic waveforms — the source-filter mechanism, in Lieberman's terminology (1975). Phonetic studies have shown that the production of human speech sounds requires the active shaping of the supralaryngeal vocal tract through movements of the tongue, velum, lips, and jaw. This would seem to require an iconic driving of the vocal tract, analogous to the iconic driving of the hand, which programs the mouth into particular shapes through temporally regulated aimed movements of the movable parts. Since spectrograms of human speech do not reveal phonemic features, it is not possible to deduce phonemic contrast from the visible sound record. However, it is possible to ask whether monkeys actively modulate sound production through supralaryngeal movement, and motion picture studies of geladas show active lip movements from lip-rounding to complete closure. It is worth recalling also Anthoney's analysis of the baboon lip smack as a tongue-alveolar ridge contact gesture — another sound class associated with face-to-face affiliative behaviors. These studies suggest vocal interactions in monkeys that have the optionality of a prestational exchange process and the semanticity implicit in morphological shape changes in the vocal tract and that are systematically related to the axes of social relationship. These studies provide a potential naturalistic model for language evolution. If one consequence of voluntary control of the vocal tract is accessibility to conceptual encoding, then similar shaping changes in vocal production could be driven by the conceptual representation of social relationships and integrated into the prestational exchange of affiliative behavior. This would give the evolutionary preconditions for social interactions that use exchanged vocalizations shaped by conceptual information about contextually relevant social categories. Moreover, since instrumental action is responsive to the state-specific recombinations of play, new vocal combinations could emerge in play behavior, as occurs now with chimpanzee signs, to enlarge the repertory of vocal productions. These monkey studies, which point toward contextual constraint on vocal productions in affiliative social situations, casts doubt on the common view of the disembodied proposition as the essence of language (Ryle 1962).

A transmission model of language, derived from culture theory, leads one to expect that the normal use of language is to pass on information from one person to another, and the concept of propositionality reinforces the view that meaning is encoded into speech and passed like a parcel to someone else. Behaviorally, however, as anthropologists have documented from various cultures, a good deal of speech says nothing at all (Malinowski 1965; Firth 1972). It is part of ritualistic greeting behavior of which the syntactic form is not intended to be taken seriously (and taking it seriously can actually be rude), such as giving an honest answer to "How do you do?" Such speech also occurs in religious contexts in which people no longer even know what the words mean. It occurs as magical charms and sacred formulas that cannot be uttered socially at all. Long hours are spent in discussion of polite topics that everyone knows about already. The most banal platitudes are uttered with portentous dignity at affairs of state. Contrary to what a transmission-propositionality model would predict, these are perfectly normal uses of human speech. There is nothing residual or deviant about them. If anything, they err in the other direction, being elaborated into ritual languages, sacred texts, and Fourth of July speeches, all of which are often treated with solemn deference. Linguistically, too, it is common knowledge that words have a connotative dimension of meaning that is as important as their denotative meaning.

The instrumental theory of language is in fact very strong at this juncture, for it would ground language firmly in its interactive setting. The instrumental theory would predict that language would be subject to all the motivations, contingencies, and intentions that characterize volitionally mediated social behavior generally. Because the affective modality of apes and humans is very similar, humans would be expected to use language as apes use their own volitional motor programs: for the negotiation of social rank, for intergroup display, in playful recombination, as instrumental aggression, as gestures of affection, and so on. However, the limitations of the instrumental theory as a general theory of language were already discussed, and any theory that aspires to general status must come to terms with the propositionality of language. Here, however, the instrumental theory could redefine the problem, and this redefinition must be taken seriously. In a consistently applied instrumental position, language is a tool that operates on the conceptual system of its perceivers. The linguistic encoding process, in other words, is intrinsically instrumental, and the difference lies in the effect of this particular tool once it reaches human ears. By this model, the uniqueness of language lies in its effects and not in its causes. This approach places linguistic reference in the brain of the perceiver and not in the utterance, and it is compatible with another definition of the proposition that would satisfy the philosophers: that propositions are behavior for which the predication of truth and falsity is meaningful. Since only concepts have that property, the instrumental theory would argue that language is an instrumentally organized form of behavior that is goal-directed to create a particular effect in the conceptual systems of its perceivers. By this analysis, the meaning of an utterance is always speaker-relative and hearer-relative. By this analysis, too, Noam Chomsky's famous sentence, "Colorless green ideas sleep furiously," is not meaningless at all but is an ingeniously constructed instrumentality designed to keep linguists talking for years. However, the instrumental theory converges at this point with the instrumental channel theory, for in asserting that language is a tool with very special effects, it must postulate a central mechanism that can mediate conceptual reorganization. We are faced once again with a special connection between language and memory.

Some of the difficulty in formulating a coherent theory of language evolution may be due to an overly sharp distinction between language and memory. It has been known for over two decades that lesions of a limbic system structure, the hippocampus, result in deficits on verbal memory tests, both visual and auditory; and this effect is lateralized to the speech hemisphere (Milner 1970). This was long interpreted as a defect in the consolidation of short-term memory into long-term memory because such patients were able to recall lists of words as long as they rehearsed them but failed after brief distractions. However, it became apparent that the defect was far more anomalous than originally suspected because about 50 percent of the intrusion errors of these patients were from word lists to which they had been exposed on previous days, indicating that although they could not recall they still had not forgotten (Isaacson and Pribram 1975). However one characterizes this impairment, it is clear that the failure to consolidate into long-term memory is not the problem. This finding removes a major discrepancy in the neuropsychological literature on the lesion results from humans and animals, because all attempts to reproduce a memory loss in animals by bilateral hippocampal lesions had failed. Now it is clear that such lesions do not produce consolidation deficits in humans either, but the exact nature of the recall problem is still not generally agreed upon. Douglas and Pribram, in an early summary of the effects of hippocampal lesions in monkeys, had concluded that this structure functions to inhibit responding by attention to "negative instances," and this formulation parallels the recall deficit because both are part of a more general inability to inhibit irrelevant response alternatives stored in memory. This approach links verbal memory to an active process of response inhibition rather than to a passive process of information retrieval. This seemingly innocent change of perspective greatly simplifies the relationship between language and memory because verbal memory is information stored in a systemic-formative code. That is, verbal memory is information stored in a motor program capable of operating the vocal tract to produce the distinctive acoustic waveforms we hear as phonemes and words. From linguistic evidence it is known that these speech programs are a system of response alternatives. Phonemes are contrastive at the level of distinctive features: English /p/ differs from /b/ only in the feature of voicing—a vibratory movement of the vocal cords; and English /b/ differs from /d/ only in place of articulation—the lips versus the alveolar ridge. In a similar manner, the basic phonemic combinations—the possible morpheme shapes of the language (as opposed to meaningful units) — are also contrastive phoneme substitutions: bat, cat, dat, rat, gat, and so on. If we think of verbal memory not as stored information but as a pattern of disinhibition (or its converse) superimposed upon a motor programming system of contrasting alternatives, then a recognition process and a motor process are not as distinct as they might at first appear. Given this relationship, it is clear that a tool that operates upon conceptualization is not impossible to conceive of, provided language is not arbitrarily divided into *language* on one hand and *memory* on the other. Language is the interlocking of motor programs with a system of memory gating that operates via the patterned disinhibition of contrasting alternatives. Language, as Freud suggested, is inherently connected to an active process of inhibition, which he termed *repression*.[12]

This analysis suggests that language is not a "kind of behavior," which can be defined in morphological terms. It is rather a central system that links motor programs of contrasting alternatives to other kinds of memory storage. However, the behavioral form of language is not arbitrary, for the instrumental channel theory would argue that the motor code used by language can be no more sophisticated than the input-output functions of instrumentality generally. For this reason, there ought to be a systematic relationship between language and instrumental action even though language is not reducible to instrumental action. In a similar way, the semantic power of language ought to be effectively constrained by the mechanisms of conceptual storage, even though language is not reducible to conceptual storage. An obvious implication of this reasoning is that language is affected by evolutionary developments of instrumentality and conceptualization, such that linguistic productions will vary with the grade of cognitive development. From the perspective of language origin theory, it should be possible to specify the potential constraints on a linguistic system through structural descriptions of other forms of constructional activity. If the structural modeling of the instrumental actions involved in material constructions proves to be possible, then there will be a potential archaeological window into the linguistic productions of extinct hominids, even though language and tool use are not the same thing. This approach also implies discontinuities in primate linguistic abilities, which parallel the differences in constructional activity already discussed. The ability to use simulated images in the constructional process should reorganize a protolinguistic system to enable the production of utterances that are more like integrated constructions than like simple lexical strings. Constructions, after all, have a derivational history that is not completely deducible from the finished product. The linguistic analog of a "derivational history not deducible from the finished product" is called a transformation rule, and Chomsky was right to insist on its phylogenetic importance. The advantage of such a system is that it permits true linguistic constructions, which are specifically designed to produce certain effects on the conceptual system of the hearer.

Although it is unsporting to change the rules on the chimpanzees at such a late stage of the game, the sequencing rules taught to apes also derive from a theory of language that emphasized the arbitrariness of syntax in a manner analogous to the arbitrariness of the lexicon. In the ingenious approach of Von Glaserfeld and Rumbaugh, a computer has been programmed with an automatic parser for an artificially devised language, Yerkish; and the chimpanzee can communicate with its trainers through sequences of key presses that are interpreted by the computer and evaluated for their conformity to the semantic and syntactic rules encoded in the parsing program (Rumbaugh 1977). However, the definition of *grammar* that has informed this experiment is "a set of conventional rules that govern the formation of sign combinations that have semantic content in addition to the meanings of the individual signs" (Von Glaserfeld 1977, p. 66). This definition emphasizes the conventionality of syntactic structures, whereas an instrumental channel theory, seriously pursued, would argue for a systematic interrelationship between syntactic structure and constructional skill. In this approach, which Von Glaserfeld in fact touches upon, grammar is not the product of convention but a product of constrained variation at a deeper cognitive level. Although culturally variable, syntax is never arbitrary and conventional to the deeper logic of its users. It must be this systematic connection between conventionalized linguistic productions and the underlying logic of skill, universally shared among human beings, that contains the answer to the paradox of the translatability of natural languages. A definition of syntax derived from progressive history will never carry us to this larger question. The constructional ability that allows linguistic productions to be tailor-made for their effect on particular audiences also entails a great deal of knowledge about what other individuals would like to hear, and this fact alone places language squarely in the domain of affective assessment. Affective action and social action are two sides of the same coin, and this indicates that the nature of the proposition, commonly defined in the denotative, empiricist terms of Enlightenment psychology, has also been generally misconstrued.

Historically, the truth value of propositions has been located in the relationship between a proposition and its referents in the external world. Behaviorally, however, a far better case could be made if the truth value of a proposition were located in the relationship between the proposition and conceptually stored information. In a behavioral analysis, language is not psychologically related to the *transmission of truth* at all but it is related to the *induction of belief*. It is only by arbitrarily segmenting the language circuit into *language* on one hand and *verbal memory* on the other that the proposition takes on this rarified existence that allows its truth value to exist independent of its ability to reorganize conceptual networks. In the functional characterization of language, it is not the "objective truth value" but the "subjective truth value" that gives language its power.

This view of language makes very different predictions from either a transmission model or a truth model. Far from expecting linguistic efficacy to correlate with the objective truth or falsity of sentences, this model would predict that language would be most efficacious at transforming conceptual networks in situations that magnified the channels of belief and the scope of innate signals: in linguistic transactions that passed from the higher ranking to the lower, in states of high positive affectivity, in utterances that move along the networks of altruistic grooming, and in utterances that can coordinate the highly affective experiences hung on the hooks of words. Truth is another metalinguistic concept that requires knowledge of both the proposition and the empirical status of its referents. Psychologically, however, at the level of actual primate behavior, a proposition always manifests itself as belief or disbelief. The conceptual reorganizations induced by language are not the transmissions of truth but the creation of belief, and language can by no means be considered an unequivocal transmission system. On the contrary, there is every reason to believe that propositional information is always gated by the ethological systems of authority and attachment that influence social information flow in other systems, like the protocultural innovations described for Japanese macaques (Tsumori 1967).

Far from being a liability, however, the close connection between language and affect is the source of all its strength. The evolutionary significance of language is due precisely to the fact that it can potentially engage the conceptual system in exactly the same way as actually perceived events. In this sense, language is a form of behavior with a reality value so heightened that its effects can be functionally equivalent to actual experience. This phenomenon suggests not the detached pattern detection of observational learning but the innate attention-getting power of species-specific signals. The crosscultural variability of language is not an argument against the innate behavioral component. In the light of ethological studies of bird song acquisition and neuropsychological demonstrations of environmental tuning of innate feature-detector cells in mammalian sensory cortex, it is clear that innate systems can interact with environmentally derived information to produce a composite mode of response that is both resistant to further change after maturation and potentially variable across generations.

As Pribram has pointed out, there is an analogy of function between embryonic induction mechanisms, as described by biologists, and the properties of reinforcement described by psychologists (1971, p. 273). In both cases, an innately significant class of stimuli interacts with the genetic potentiality of undifferentiated tissue to induce a specific morphology through processes of development. Although learning is often considered to be one kind of phenomenon and growth another, reinforcement may in fact be the embryonic induction that continues to function in the mature brain. In labile behavioral systems, the distinction between induction and reinforcement may in fact be meaningless. From this perspective, language is an embryonic induction system that locks lexical units, based on a systemic-formative code of contrasting alternatives, into a developing conceptual network in the course of maturation. If language is considered such a process of postnatal growth, which merges reinforcement and memory, theories of language acquisition are not likely to be successful if they simply focus on general cognition and ignore the magic of words. A far more ethological theory of language development may be called for: one which specifies a central circuit capable of interrelating the mechanisms of context-free epicritic processing to enduring changes in conceptually stored information through engagement of the innate attention–getting power of social signals. Such an ethological theory may not be as far in the future as might be supposed. If language is viewed as an induction process that can transform conceptual networks, then there is a functional similarity between language and two other systems already singled out as behaviorally significant in primates: play and active (REM) sleep. The similarities of these two states, first noted by Gregory Bateson, were explained as state-specific reorganizations of sensory and motor networks. Special attention was drawn to the productive nature of behavioral combinations in these states, and productivity is one of the significant defining features of language. Like the postulated functions of play and active sleep, language also has reorganizing properties, but the reorganization is not in the maturation of the behavioral networks but in the alteration of an existing conceptual store. As noted also, language has extreme psychological saliency, which engages the attentional mechanisms of the human brain with the force of species-specific signals. Both of these properties argue for the close functional involvement of the limbic system in linguistic processing. Since limbic system structures are known to be involved in regulating developmental metabolism through their sensitivity to circulating hormones, which has been well documented in the case of a loop connecting the hippocampus to hypothalmus to pituitary to ACTH to adrenal cortex to corticosteroid hormones and back to the hippocampus (Isaacson and Pribram 1975; Fuxe et al. 1973; Wied and Gispen 1977; Kernadakis 1971), a biological approach to language would try to define a central mechanism that could relate cortical detection of species-specific signals to enduring changes in the neural function of adult brains. From this perspective, language is the coupling of instrumental behavior to species-specific vocal signals on one hand and to a fast-time induction system on the other.

SUMMARY

The hypothesis that language is a form of tool using (the instrumental theory) is examined and rejected on the grounds that it does not adequately distinguish the peculiarities of language from other forms of rule-governed, productively generated, intentional action. However, the claim that language is based on an emergent symbolic faculty for conventional and arbitrary equivalences does not take account of response-contingent learning in language acquisition, the symbolic performances of apes, the close relationship between language skills and cognitive skills generally, nor the "naturalness" of a language to native speakers. Instead a coevolutionary model is adopted, in which language is theorized to evolve in conjunction with the conceptual and motor advances of the instrumental modality and to use instrumental action as an input–output device but nonetheless to remain a distinct system. The psychological prototype of language is seen as an inhibitory grid superimposed on an innate motor network of contrasting vocal productions, which can be selectively and partially activated by conceptually stored contextual information. The patterned disinhibition of vocal productions can thus serve as a motor memory, and psychological research indicates a close functional connection between the language code and human memory. Since the vocal motor programs involved are jointly controlled by both the instrumental and affective modalities, they make possible a new composite behavioral product, free from the "real world" constraints that operate in the manipulation of matter, that can be specifically evolved for its effect on listeners. These composite vocal productions, jointly produced by affect and instrumentality, can simultaneously activate the mechanisms of affective information processing with all the attention-getting power of innate signals and engage the interpretive mechanisms of conceptual (epicritic) processing with the most advanced instrumental programs of the primate brain. The social prototype of language, suggested by studies of Japanese macaques, is seen as involving (1) volitional control of innate vocal behaviors during mutual vocal exchange, such as the prestational interactions of social grooming, and (2) the modulation of these vocal behaviors by conceptually stored information that differentiates the ongoing context from contrasting alternatives. Such volitionally controlled but composite vocal productions ought to become possible with the phylogenetic integration of the vocal tract into the instrumental modality, and the vocal productions themselves should vary discontinuously in the course of evolution in response to discontinuities in constructional action. On the receptive side, the psychological precursor of language is hypothesized to be a fast-time embryonic induction circuit, which locks perceived vocal productions into the inhibitory grid during maturation in children and selectively activates the development of conceptual connectivity in adults. The affective component of human speech allows these motor programs and their external acoustic products to engage the human brain with the force of an embryonic inductor, and this in turn argues for close links between language and affect, both in the face-to-face contexts in which speech develops and in the match between the morphology of utterances and the requirements of affective "deep structure."

To summarize the summary, language is a system of affective-instrumental integration, which can produce composite behavioral constructions capable of operating as a gate on conceptually stored information of listeners through the selective disinhibition of contrasting motor programs that are themselves connected to memory storage. Language is indeed an evolutionary emergent, but it requires a new integration of preexisting components and not the elaboration of new principles.

NOTES TO CHAPTER 6

1. See also Menzel and Halperin (1975).

2. See McDermott and Roth (1978) for anthropological studies using this perspective.

3. See chap. 4.

4. Fouts et al. (1976) give evidence of vocal understanding in an ape.

5. See chap. 1.

6. See also chap. 4 for cross-modal transfer tasks.

7. See also Geschwind's discussion of Myers (in Myers 1967, p. 72).

8. Cheremic contrast is discussed by Stokoe (1975).

9. See Lieberman (1975) for explication, also Fry (1977).

10. See Dewson and Cowey references in chap. 4.

11. Also see references by Sperry and Levy in chap. 3.

12. For Freud, repression was a failure to attend to certain stored events associated with a word. It is interesting in this respect that recent approaches to computerized comprehension of human speech have used the opposite of repression, namely selective attention, as a major linguistic construct (Grosz 1977).

REFERENCES

BELL, D. S. 1968. Speech functions of the thalamus inferred from the effects of thalamotomy. *Brain* 91:619–638.

BELLUGI, URSULA, and KLIMA, EDWARD S. 1975. Aspects of sign language and its structure. In *The role of speech in language,* ed. J. F. Kavanagh and J. E. Cutting, pp. 171–203. Cambridge: MIT Press.

1976. Two faces of sign: iconic and abstract. *Annals N.Y. Acad. Science* 280:514–538.

BELLUGI, URSULA; KLIMA, EDWARD S.; and SIPLE, PATRICIA 1975. Remembering in signs. *Cognition* 3, no. 2: 93–125.

BERLIN, CHARLES I. 1977. Hemispheric asymmetry in auditory tasks. In *Lateralization in the nervous system,* ed. S. Harnad, pp. 303–323. New York and London: Academic Press.

BOGEN, J. E., and BOGEN, G. M. 1976. Wernicke's region—where is it? *Annals N.Y. Acad. Science* 280:834–843.

BRONOWSKI, J., and BELLUGI, URSULA 1970. Language, name and concept. *Science* 168:669–673.

BRUNER, JEROME 1973. Organization of early skilled action. *Child Development* 44:1–11.

CHAPMAN, ROBERT M., et al. 1977. Semantic meaning of words and average evoked potential. In *Language and hemispheric specialization; cerebral event-related potentials,* ed. J. E. Desmedt, pp. 36–47. Basel: Karger.

DOUGLAS, MARY 1972. Deciphering a meal. *Daedalus* (winter). 61–81.

FRY, DENNIS 1977. *Homo loquens: man as a talking animal.* Cambridge: Cambridge U. Press.

FIRTH, RAYMOND 1972. Verbal and bodily rituals of greeting and parting. In *The interpretation of ritual. Essays in honor of A. I. Richards,* ed. J. S. La Fontaine, pp. 1–38. London: Tavistock.

FOUTS, R. S.; CHOWN, B.; and GOODIN, L. 1976. Transfer of signed responses in American sign language from vocal English stimuli to physical object stimuli by a chimpanzee (*Pan*). *Learning and Motivation* 7:458–475.

FREUD, SIGMUND 1953. *On aphasia: a critical study.* Trans. E. Stengel, pp. 44–105. New York: International Universities Press. Reprinted in *Brain and Behaviour,* ed. K. H. Pribram, vol. 4, pp. 13–57. Harmondsworth, Eng.: Penguin. German ed., 1901.

    1975. *The psychopathology of everyday life.* Trans. Alan Tyson. Harmondsworth, Eng.: Penguin. 1st German ed., 1901.

FRIGYESI, T. L.; RINVIK, ERIC; and YAHR, M. D. eds. 1972. *Corticothalamic projections and sensorimotor activities.* New York: Raven Press.

FUXE, K.; HOKFELT, T.; JONSSON, G.; and LIDBRINK, P. 1973. Brain-endocrine interaction: are some effects of ACTH and adrenocortical hormones on neuroendocrine regulation and behaviour mediated via central catecholamine neurons? In *Hormones and brain function,* ed. K. Lissak, pp. 409–425. New York: Plenum Press.

GESCHWIND, NORMAN 1964. The development of the brain and the evolution of language. *Monograph Series on Language and Linguistics* 17:155–169.

    1965. Disconnexion syndromes in animals and man. *Brain* 88, pt. 1:237–294; pt. 2:585–644.

    1970. The organization of language and the brain. *Science* 170:940–944.

GOFFMAN, ERVING 1959. *The presentation of self in everyday life.* Garden City, N.Y.: Doubleday Anchor Books.

    1961. *Encounters: two studies in the sociology of interaction.* Indianapolis, Ind.: Bobbs-Merrill.

    1975. *Frame analysis: an essay on the organization of experience.* Middlesex, Eng.: Penguin. 1st ed., Harper and Row, 1974.

GOODSON, BARBARA DILLON, and GREENFIELD, PATRICIA MARKS 1975. The search for structural principles in children's manipulative play: a parallel with linguistic development. *Child Development* 46:734–746.

GREEN, STEVEN 1975a. Variation of vocal pattern with social situation in the Japanese monkey *(Macaca fuscata):* a field study. In *Primate behavior: developments in field and laboratory research,* ed. Leonard Rosenblum, vol. 4, pp. 1–102. New York: Academic Press.

    1975b. Dialects in Japanese monkeys: vocal learning and cultural transmission of locale-specific vocal behavior. *Z. Tierpsychol.* 38:304–314.

GREENFIELD, P. M.; NELSON, K.; and SALTZMAN, E. 1972. The development of rulebound strategies for manipulating seriated cups: a parallel between action and grammar. *Cognitive Psychol.* 3:291–310.

GREENFIELD, PATRICIA MARKS, and SMITH, JOSHUA H. 1976. *The structure of communication in early language development.* New York and London: Academic Press.

GROSZ, BARBARA J. 1977. The representation and use of focus in dialogue understanding. *SRI International Technical Note* no. 51.

HAMILTON, J. 1974. Hominid divergence and speech evolution. *J. Human Evolution* 3:417–424.

HEWES, GORDON P. 1973a. Primate communication and the gestural origin of language. *Current Anthropology* 14, nos. 1–2: 5–32.

1973b. An explicit formulation of the relationship between tool-using, tool-making, and the emergence of language. *Visible Language* 7, no. 2: 101–127.

HOCKETT, CHARLES F. 1960. The origin of speech. *Scientific American* 203, Sept.:88–111.

HOCKETT, CHARLES F., and ASCHER, ROBERT 1964. The human revolution. *Current Anthropology* 5:135–168.

HOLLIEN, HARRY 1975. Neural control of the speech mechanism. In *The nervous system,* ed. D. B. Tower, vol. 3, pp. 483–491. New York: Raven Press.

HUTTENLOCHER, JANELLEN; EISENBERG, K.; and STRAUSS, SUSAN 1968. Comprehension: relation between perceived actor and logical subject. *J. Verbal Learning and Verbal Behavior* 7:527–530.

ISAACSON, R. L., and PRIBRAM, K. H., eds. 1975. *The hippocampus.* 2 vols. New York and London: Plenum Press.

JAKOBSON, ROMAN 1964. Towards a linguistic theory of aphasic impairments. In *Ciba Foundation Symposium on Disorders of Language,* ed. A. V. S. de Reuck and M. O'Connor, pp. 21–46. London: Churchill.

JAKOBSON, ROMAN, and HALLE, MORRIS 1956. *Fundamentals of language.* The Hague: Mouton.

KENDON, ADAM 1975. Gesticulation, speech, and the gesture theory of language origins. *Sign Language Studies* 9:349–373.

KIMURA, D., and ARCHIBALD, Y. 1974. Motor functions of the left hemisphere. *Brain* 97:337–350.

KORNHUBER, HANS H. 1977. A reconsideration of the cortical and subcortical mechanisms involved in speech and aphasia. In *Language and hemispheric specialization in man: cerebral event-related potentials,* ed. E. J. Desmedt, pp. 28–35. Basel: Karger.

LENNEBERG, ERIC 1960. Language, evolution, and purposive behavior. In *Culture in history: essays in honor of Paul Radin,* ed. Stanley Diamond, pp. 869–893. New York: Columbia U. Press.

1962. Understanding language without ability to speak: a case report. *J. Abnormal and Social Psychol.* 65, no. 6:419–25.

1964. Speech as a motor skill with special reference to nonphasic disorders. In *The acquisition of language,* ed. U. Bellugi and R. W. Brown, Serial no. 92, vol. 29, pp. 115–127. Monograph, Society for Research in Child Development.

LIBERMAN, A. M. 1970. The grammars of speech and language. *Cognitive Psychol.* 1:301–323.

LIBERMAN, A. M.; COOPER, F. S.; SHANKWEILER, D. P.; and STUDDERT-KENNEDY, M. 1967. Perception of the speech code. *Psychological Review* 74:431–461.

LIEBERMAN, P. 1975. *On the origins of language: an introduction to the evolution of human speech.* New York: Macmillan.

LURIA, A. R. 1964. Factors and forms of aphasia. In *Ciba Foundation Symposium on Disorders of Language,* ed. A. V. S. de Reuck and M. O'Connor, pp. 143–161. London: Churchill.

McDERMOTT, R. P., and ROTH, DAVID R. 1978. The social organization of behavior: interactional approaches. *Ann. Rev. Anthropology* 7:321–345.

MALINOWSKI, B. 1965. *Coral gardens and their magic.* vol. 2. Bloomington: Indiana U. Press. 1st pub., 1935.

MENZEL, E. W. 1972. Spontaneous invention of ladders in a group of young chimpanzees. *Folia Primatol.* 17:87–106.

MENZEL, EMIL W., and HALPERIN, STEWART 1975. Purposive behavior as a basis for objective communication between chimpanzees. *Science* 189:652–654.

MILNER, BRENDA 1970. Memory and the medial temporal regions of the brain. In *Biology of memory,* ed. K. H. Pribram and D. E. Broadbent, pp. 29–50. New York and London: Academic Press.

MYERS, RONALD E. 1967. Cerebral connectionism and brain function. In *Brain mechanisms underlying speech and language,* ed. C. H. Millikan and F. L. Darley, pp. 61–72. New York and London: Grune and Stratton.

NEWMAN, J. D., and SYMMES, D. 1974. Arousal effects on unit responsiveness to vocalizations in squirrel monkey auditory cortex. *Brain Research* 78:125–138.

NOTTEBOHM, F. 1970. Ontogeny of bird song. *Science* 167:950–956.

OJEMANN, G. A., and WARD, A. A., Jr. 1971. Speech representation in ventrolateral thalamus. *Brain* 94:669–680.

OSGOOD, CHARLES E. 1964. Semantic differential techniques in the comparative study of cultures. *Amer. Anthropologist* 66:171–200.

OSGOOD, CHARLES E.; MAY, WILLIAM H.; and MIRON, MURRAY S. 1975. *Cross-cultural universals of affective meaning.* Urbana: U. of Illinois Press.

PALERMO, DAVID S. 1975. Developmental aspects of speech perception: problems for a motor theory. In *The role of speech in language,* ed. James F. Kavanagh and James E. Cutting, pp. 149–154. Cambridge, Mass.: MIT Press.

PENFIELD, WILDER, and ROBERTS, LAMAR 1959. *Speech and brain-mechanisms.* Princeton, N.J.: Princeton U. Press.

PREMACK, DAVID, and WOODRUFF, GUY forthcoming. Does the chimpanzee have a theory of mind? *The behavioral and brain sciences.*

PRIBRAM, KARL H. 1971. *Languages of the brain.* Englewood Cliffs, N.J.: Prentice-Hall.
1972. Association: cortico-cortical and/or cortico-subcortical. In *Cortico-*

thalamic projections and sensorimotor activities, ed. Tamas L. Frigyesi, Eric Rinvik, and Melvin D. Yahr, pp. 525–549. New York: Raven Press.

RATNER, NANCY, and BRUNER, JEROME   1978. Games, social exchange and the acquisition of language. *J. Child Language* 5:391–401.

REYNOLDS, PETER C.   1976. Language as a skilled activity. *Proc. N.Y. Acad. Science* 280:150–166.

    Forthcoming. The primate constructional system: The theory and description of instrumental object use in humans and chimpanzees. In *Approaches to the study of goal directed action.* ed. M. Von Cranach and R. Harré. Cambridge: Cambridge U. Press.

RICHMAN, B.   1976. Some vocal distinctive features used by gelada monkeys. *J. Acoust. Soc. Amer.* 60:718–724.

RIKLAN, MANUEL, and COOPER, IRVING S.   1977. Thalamic lateralization of psychological functions: psychometric studies. In *Lateralization in the nervous system,* ed. S. Harnad et al., pp. 123–133. New York and London: Academic Press.

RUMBAUGH, DUANE M.   1977. *Language learning by a chimpanzee. The LANA project.* New York: Academic Press.

RYLE, GILBERT   1962. The theory of meaning. In *The importance of language,* ed. Max Black, pp. 147–169. Ithaca and London: Cornell U. Press. 1st pub., 1957.

SAVAGE-RUMBAUGH, E. SUE; RUMBAUGH, DUANE M.; and BOYSEN, SALLY forthcoming. Linguistically mediated tool use and exchange by chimpanzees *(Pan troglodytes). The behavioral and brain sciences.*

SAVAGE-RUMBAUGH, E. SUE; WILKERSON, B. J.; and BAKEMAN, R.   1977. Spontaneous gestural communication among conspecifics in the pygmy chimpanzee. *(Pan paniscus).* In *Progress in ape research,* ed. G. H. Bourne, pp. 97–116. New York and London: Academic Press.

SCHWARTZ, GARY E.; DAVIDSON, RICHARD J.; and MAER, FOSTER   1975. Right hemisphere lateralization for emotion in the human brain: interactions with cognition. *Science* 190:286–288.

STEKLIS, HORST D., and HARNAD, STEVAN R.   1976. From hand to mouth: some critical stages in the evolution of language. *Annals N.Y. Acad. Science* 280:445–455.

STOKOE, WILLIAM C., JR.   1975. The shape of soundless language. In *The role of speech in language,* ed. J. F. Kavanagh and J. E. Cutting. Cambridge, Mass.: MIT Press.

TERRACE, H. S.   1979. How Nim Chimsky changed my mind. *Psychol. Today,* Nov., pp. 65–76.

THOMPSON, R. K. R., and HERMAN, L. M.   1977. Memory for lists of sounds by the bottle-nosed dolphin; convergence of memory processes with humans? Science 195:501–503.

TSUMORI, ATSUO 1967. Newly acquired behavior and social interactions of Japanese monkeys. In *Social communication among primates,* ed. Stuart Altmann, pp. 207–219. Chicago: U. of Chicago Press.

VERNADAKIS, ANTONIA 1971. Hormonal factors in the proliferation of glia cells in culture. In *Influence of hormones on the nervous system,* ed. D. H. Ford, pp. 42–55. Basel: Karger.

VON GLASERFELD, ERNST 1977. Linguistic communication: theory and definition. In *Language learning by a chimpanzee. The LANA project,* ed. D. M. Rumbaugh, pp. 55–71. New York and London: Academic Press.

WARREN, RICHARD M. 1976. Auditory perception and speech evolution. *Annals N.Y. Acad. Science* 280:708–717.

WATERS, R. S., and WILSON, W. A. 1976. Speech perception by rhesus monkeys: voicing distinction in synthesized labial and velar stop consonants. *Perception and Psychophysics* 19:285–289.

WEBSTER, DOUGLAS, and WEBSTER, MOLLY 1974. *Comparative vertebrate morphology.* New York and London: Academic Press.

WIED, D. DE, and GISPEN, W. H. 1977. Behavioral effects of peptides. *Peptides in neurobiology,* ed. Harold Gaines, pp. 397–448. New York and London: Plenum Press.

ZANGWILL, O. L. 1975. Excision of Broca's area without persistent aphasia. In *Cerebral localization,* ed. K. J. Zulch et al., pp. 258–263. Berlin: Springer Verlag.

7

IN QUEST OF CONTINUITY

CONTINUITY PHENOMENA

The evidence reviewed in the preceding chapters indicates that there are both quantitative and qualitative behavioral changes involved in the evolutionary transition from apes to hominids. Some of these changes can be classed as *continuity phenomena* in that they can be modeled by evolutionary trends and processes observable in nonhuman primates or other animals. Many of the progressive aspects of human cognition can be postulated to be related to increases in the volume of the brain and the reorganizational consequences following from that. Since similar transitions are observable among contemporary primates, these phylogenetic trends can in principle be modeled by cognitive differences in other species. Among these would be the progressive differentiation of the body image (revealed in motor control), the incorporation of external objects into motor programs, and the increased capacity for body image incorporation revealed by Gallup's experiments.[1] These differences are likely to reflect an evolutionary progression in the motor and parietal cortex, as these areas are involved in kinesthetic and body image processing in man, and similar changes occur developmentally.[2] The *distalward migration of effectors* is also observable in primates in the progression from mouth to hand to tonic grip to constructional manipulation. As Hershkovitz has documented, a similar progression is also observable in the affective signals (1977). The regions of the primate body that develop distinctive colors and hair patterns, used in signaling, are also the regions that correspond to the tactile vibrissal fields of primitive mammals. This implies that the distance receptors of higher primates, as adults, are mediating affective information that was once carried in the tactile modality. From a neurological perspective, it argues for increased visual input into the limbic system in the course of phylogeny. Again, this is a sensory modality shift and not attenuation of the central connections of the vibrissal fields. We can regard some of the iconic constructions of humans as the culmination of such a trend. There is also a related trend toward the *iconic driving* of distal effectors, suggested by the central image capacities of monkeys, the sign hand shapes of apes, and the iconic constructional abilities of humans. Some features of this process can be studied in nonhuman primates. There are also likely to be qualitative performance differences, which reflect quantitative differences in the capacity of the conceptual store to merge information or deal with a given depth of nesting. These, too, can be studied among existing primates.

Many of the differences between human and ape modes of social organization and subsistence, postulated as significant by anthropological investigation, can also be accommodated within an existing theoretical framework. Isaac has argued that hominids form an adaptive complex consisting of *cooperative hunting*, a *home base, foraging parties*, a *sexual division of labor, food exchange*, and *tool making* (1976, 1978). An examination of primatological data reveals that humans differ from all other primates in this array of features, but the changes involved are no greater than those that have occurred among other species. Hunting has prototypes already considered, and cooperative hunting occurs in rudimentary form in chimpanzees. The extensive mammoth kills of the Middle Pleistocene, well documented archaeologically, argue for much more sophisticated cooperation than any seen in existing apes (Howell 1965) although the interaction processes by which human cooperation is organized are only now getting the attention of anthropologists (McDermott et al. 1978). However, apes provide a model of this development, and their cooperative activity has been studied experimentally (Crook 1971; Crawford 1937, 1941; Nissen and Crawford 1936).[3] Also, evolution of the body image would allow a further refinement of observational learning that would allow cooperation and coordination through the inferred intention of another's action, making possible genuine technology.

The emergence of a home base, to which foraging parties return at night, inferred by Isaac in his archaeological investigations of East African hominid sites, is also a continuity phenomenon (1975). A social organization of group fission and fusion, sustained by kinship and age-mate attachment relationships, male status relationships, and sexual attraction, already occurs in chimpanzees (Reynolds and Reynolds 1965). The daily partings and joinings of subgroups within a common community, documented by Nishida, can be thought of as a temporal and spatial extension of processes that are also observable in the multimale troops of baboons and macaques on a much smaller scale (Nishida 1968). The greater cognitive component of hominid attachment systems creates the possibility for such subgroup dynamics to be extended in space and time in a way not possible with monkeys. This interpretation is consistent with the elaboration of greeting behavior in chimpanzees and with the long-term social relationships between male chimpanzees and their mothers, which indicate the reactivation of existing bonds (Lawick-Goodall 1975). The home base itself is commonplace in the animal world, and it is likely that its emergence reflects the presence of male care giving, because that would provide a strong motivation to return to the location of females and young by males.

The evolutionary importance of hominid pair-bonding has been much emphasized, but it is best construed as an increase of male-female care giving and more exclusive sexual relationships within a social organization similar to a multimale troop.[4] One compelling reason for this hypothesis is that the integration of multiple adult males in the same social organization is problematical in primates, since the vast majority of species live in exclusive mated-pair troops, age-graded multimale troops with one breeding male, or in harems. It would appear to be far easier to further differentiate an existing multimale troop, in which cooperative relationships among unrelated adult males already exist, than to create such relationships by aggregating males with exclusive breeding units. Also, the hominid bond does not have the exclusivity of truly pair-bonded species, and such exclusive bonds with their attendant same-sex jealousy would be incompatible with cooperative adult groups. The human pair bond is also very variable in duration and exclusivity, both crossculturally and among individuals, and adultery and divorce give every indication of being as venerable as marriage. Lastly, the sexual dimorphism of body size in hominids suggests a pair bond where differences in rank are of de facto importance in the genesis of the bond and where sexual selection has an opportunity to occur (Crook 1972; Trivers 1972). This, too, argues for a multimale troop. The evolutionary prototype of the hominid troop is not the nuclear family of gibbons, however natural that may seem to Westerners, but the prolonged grooming relationships and preferred sexual partners that have been documented in the multimale troops of chimpanzees, macaques, and baboons. The social organization of these societies is multilevel, with one individual participating in a number of subgroups — ranking system, peer group, lineage, coalitions, sexual dyad, and so on. This simultaneous membership in multiple groups is universally human, whereas atomistic nuclear families on defended territories occur nowhere among humans as the highest integrative level. Multimale troops can also integrate unrelated adult males (at least among monkeys), and males can form cooperative male relationships (at least among chimpanzees). Consequently, it would be straightforward to transform them in a hominid direction by increased male care of females and young. The concept of the instrumental modality considerably simplifies this transition. All that is required is the further contextualization of sexual cues, the incorporation of object-using programs into preexisting grooming networks, and further selection for male care giving by increasing grooming. Since male-female and male-child attachment relationships would already have a grooming component, it would need only to become "objectified" with the distalward migration of instrumentality, such that the primitive instrumental act of grooming is supplemented by the advanced instrumental act of object exchange.

The attachment relationships themselves are continuity phenomena and do not require a major departure from current ethological theory (Reynolds 1976). Continued sexual receptivity in females has been implicated in this development, but sexuality by itself does not provide an attachment mechanism. This is only achieved by the integration of sex with affectivity. That sex has a role, however, is consistent with a loss of direct hormonal control of many aspects of behavior. Postconception estrus cycles are common in macaques, and anestral sexual receptivity has been reported for chimpanzees (Loy 1971; Hanby et al. 1971; Lemon and Allen 1978). The development of permanent morphological indicators of sexual maturity in humans to replace cyclical indicators of receptivity, if such there were, is also consistent with this trend. The relative childlike physical attributes of sexually mature human females compared to males—smaller size, the eye size—head size ratio, subcutaneous fat, relative hairlessness of body—argue for a male care-giving releaser function in female morphology (Eibl-Eibesfeldt 1971). Also, the pendulous breasts of females (of some races and physiques) and large penis size of males also appear to be permanent sexual displays. The human female breast, unique among primates, also mimics the eye-spot pattern which is consistent with the fear-affiliation role of the latter. This interpretation is supported by the development of chest display in gelada monkeys of both sexes, who forage in a sitting position, which mimics the female genitalia (Wickler 1967). The convergence of bearded faces of adult males with the ventral body surface of adult females is also noteworthy. Since all of these morphological changes are congruent with each other, it seems that nature has spared no expense in keeping male and female hominids in proximity to one another with high levels of mutual affectivity. This argues for a *multiloop* hominid pair bond, in which a number of processes can cycle concurrently and reciprocally influence each other. Given the factors already discussed, a number of possible dyadic loops could function together to create the hominid bond. It is important to stress that these dynamics would not be derivable from introspection, because only their effects, in the form of feelings, would be subjectively accessible.

With the female as inducer, the following pathways are likely: (1) a female-to-male induction of inhibited escape via the breast eye-spot configuration (infantile dependency); (2) a female-to-male induction of care giving via the quasi-infantile morphology; (3) a female-to-male induction of sexual motivation via sexual releasers (perhaps the same as loops one and two); (4) a female-to-male induction of aggression via a secondary pathway from male inhibited escape; (5) a female-to-male induction of play also via a secondary pathway from male inhibited escape. With the exception of loop two, all mechanisms can also occur in a male-to-female version, but there would be differences because of sexual asymmetries. The larger body size and erect penis in the male would substitute for the eye-spot in loop one, and a sex difference in grooming threshold would provoke more female displacement care giving in the same conflict situation. Similarly, a sex difference in fear thresholds would lead to approach-avoidance differences in the two sexes, which would show up as differences in the affective modulation roles of the participants, namely, the females controlling approach. The induction of care giving in females might also be performed by the male facial configuration and by a secondary pathway from his response to loop one, since inhibited escape is characteristic of males in the mother-child bond and could be expected to evoke care giving in females. These postulated interrelationships are in accord with the roles of body size, infantile morphology, and sexual signals that have been reported for other species of primates. However, the signal morphology itself shows species-specific features, and the affective modality differs in hominids by virtue of hyperthresholds and its access to instrumentality. The model also predicts a relationship between sexuality and infantile dependency, as Freud observed, and it also predicts that sexual attractiveness will decrease in adulthood because the "parent" configuration will dominate the "infantile" configuration in somatic morphology.[5]

The altered role of kinship relations in hominids also appears to be a continuity phenomenon in many respects (Fox 1975). If the social significance of kinship is to organize the web of social relationships by extending and contracting the group boundaries, then this process requires the attachment systems to supply its motivating power. Yet if kinship were coextensive with the attachment boundaries, it would be superfluous. Kinship, like prestational exchange, would be expected to occupy the sphere between attachment and the outer margins of familiarity. It presupposes both the attachment systems in the service of instrumentality and the development of prestational exchange networks that use attachment figures as tokens in the exchange. As Fox points out, both de facto descent and alliance exist in nonhuman primate societies but not as an integrated system. Anthropologists have emphasized that human kinship systems are not simply ways of reckoning genealogical relationships but are ways of regulating mating and creating alliances between social groups. A potential evolutionary precursor of kinship has in fact been described for nonhuman primates. The agonistic buffering of *Macaca sylvanus*, with the prestation of infants among males, provides a nonhuman analog to the sister exchange regarded by Levi-Strauss as the prototypical kinship act (Deag and Crook 1971). In either case, adult males are using lower-ranking individuals as exchange tokens in negotiated social relationships. If the kind of behavior observed in

Macaca sylvanus occurred in early hominid troops in the context of extended male–female pair bonding, then adult males would have access to their own offspring as tokens in male exchange relationships. Since it is males that are prominent in nonhuman primate intergroup relationships, at least among sexually dimorphic species (see following discussion), a first step in a hominid level of kinship may have been intergroup prestational exchange among adult males using lower-ranking, nonmated attachment figures as tokens. It is interesting too that language is not a prerequisite to the formation of social relationships through the prestational exchange of attachment figures, though language extends the process by extending the range of "relative" beyond the boundaries of actual attachment. The prestational use of attachment can be seen as the evolutionary convergence of descent and alliance, and extended male care giving is a likely precursor.

The sexual division of labor also exhibits continuity. Since instrumental action and cooperative relationships are widespread in primates, even though nonhuman primate subsistence activity is largely individualistic, it is easier to incorporate subsistence skills into an already existing *social* division of labor than to derive the division of labor from sex differences in subsistence. From this perspective, the noneconomic social division of labor is primary and the economic division of labor, of male hunters and female gatherers, is derivative. Studies of nonhuman primates show that intergroup relations are usually male roles in species dimorphic for size.[6] Since all-female troops with female alpha animals have been reported in the wild for monkeys for short periods of time, this effect may be due to male specialization for intergroup display and the inhibition of females in the presence of males (Neville 1968). The mutual inhibitory effects of males and females have already been noted in Goldfoot's experiments with rhesus monkeys, and the role of high-ranking animals in preventing *intra*group agonism has also been experimentally demonstrated in macaques by several investigators (Oswald and Erwin 1976; Dazey et al. 1974; Hendricks et al. 1975; Tokuda and Jensen 1968; Bernstein et al. 1974; Bernstein 1964*b*; Sackett et al. 1975; Steklis and King 1978). These studies demonstrate a systematic relationship between social rank and the functions of intergroup relations and intragroup social control. As already discussed, high-ranking individuals are not usually the most aggressive, but their participation or withdrawal has a disproportionate effect on the outcome of intergroup encounters, providing good evidence for a social center and periphery in multimale troops. The transition to the cooperative hunting of *large* game (animals over fifteen kilograms), which is another hominid specialty, may have been partly mediated through the cooperative agonism of adult male hominids in response to external threats to the group. This suggests that hominid cooperative hunting of large game may be psychologically quite different from opportunistic predation of small animals by contemporary nonhuman primates. Rather, it would more closely resemble the intra-specific agonism at the margins of chimpanzee ranges, and systematic cannibalism and large game hunting should be coemergent.

The emergence of tool making and tool carrying also has strong continuity aspects. The concept of the instrumental modality maintains that tool use, in a psychological sense, antedates the incorporation of objects into the modality. However, it also argues that a hominid level of tool use is a form of observational learning and not a transmission of material objects from one generation to another. Since a *tool* is the pattern of conceptual relationships and motor organization that allows an object to be used in a goal-directed manner, it is clear that only material objects are directly transmitted from one generation to the next. Tools are never transmitted. They are rediscovered each generation by the perception of the consequences of action. What evolves in hominids is not a capacity for culture at all but an increased capacity to infer the consequences of action from the observation of others. In this sense, too, the psychological precursors to tool making are present in apes, but apes are deficient in the iconic driving of the constructional process with simulated images, as indicated above. Thus, there are aspects of the constructional process that are genuine *discontinuity* phenomena. In addition to this, as the chapter on language indicated, humans possess a system of communication without parallel in any other species of primate.

DISCONTINUITY PHENOMENA

Human technology and language, the two discontinuities stressed by the Enlightenment, are as good characterizations of the collective activities of the human species now as when they were first articulated. Yet it would be an error to conclude that an eighteenth century conception of the relationship between man and animal is therefore vindicated. Although language and technology are new emergents on the primatological landscape, the social thought of the eighteenth century was content merely to note that fact, whereas evolution has taught us to go beyond the fully formed appearances of things to continuities of process that create and sustain transformations from their more primitive antecedents. Small changes can have big effects, and large differences in functional capacity can emerge from minor alterations in the underlying morphology. An evolutionary approach to human behavior should begin with this premise and look for the new functional significance of rather minor reorganizations of existing structures.

In the discussion of ethological theory in chapter 2, a problem was raised and an answer hinted at: the curious convergence in the cultural products of man with the innate signals of animals, such as phallic sculpture and penile displays. It was noted that classical ethology and culture theory are both incapable of dealing with that juxtaposition, for it requires the integration of advanced object constructional activities with the innate perceptual components of affectivity. In the discussion of language in chapter 6, a similar argument was put forth, in a different context, on the hominid propensity to produce composite behaviors, jointly controlled by the affective and instrumental modalities. What has not been discussed is the evolutionary potential of this development. If hominids are producing novel behavioral productions, under the impetus of conceptual information and affective need, which simulate features of innate signals, then it follows that constructed signals can activate the human instinctive mechanisms that perceive those features. That is, an entirely new channel of social coordination emerges in hominids, but it is based on phylogenetic continuity in affective perception conjoined to phylogenetic discontinuity in affective transmission. Moreover, the old channel, with continuity in both affective perception and transmission, as shown by facial expression, also continues in its own sphere of specialization. In other words, *cultural integration in man, although clearly a new functional system, is based on the persistence of instinct*, not only in affective perception and motivation, but in conceptual encoding of innately determined relationships. Without instinct, humans would not only behave differently, as would be expected, but the content of their conceptually stored information would be dramatically altered as well.

In a weak version of this formulation, hominids simply construct simulations of innate features that map directly onto innate perceptual mechanisms, like manufacturing baby dolls or painting eye spots on a shield. Although hominids clearly do this, such activity is probably only a minor fraction of processes of affective-instrumental integration far more involuted than anything yet unearthed by modern ethology. As noted in chapters 2 and 3, the innate motor and perceptual components do not exist in isolation but are interconnected to yield complex interrelationships. Maternal behavior, for example, not only includes the acts of care giving but is systematically connected with aggression to provide defense of the young. Similarly, in its constructional aspects, affective control produces complex products, in which features of a number of innate components can be present simultaneously. The reader need only reflect on the complex condensed symbols of dreams, delineated by Freud, or peruse the baroque accretions of symbols of the world's religious traditions to appreciate this fact. Such condensed and syncretic constructions are so widespread among humans as to be typical, and they argue against a simple externalization and activation of single innate perceptual modules. Rather, they suggest that composite images are being created to reorganize, in complex ways, the connectivity of the affective system, using the networks of memory-encoded associations. Since the constructions created through instrumental object use endure in space and time, humans can develop new mechanisms of social coordination, leading to a greatly different scale of institutional action, simply by instrumentally creating and controlling a technology of affect, *viz.,* those composite images most efficacious in transforming the affective networks of perceivers in the desired direction. The discontinuity of effect is profound, but the channel of social coordination is nonetheless the instrumental control of affectivity through progressive image-driven constructional skills conjoined to phylogenetically conservative perceptual modules. However repugnant the thought may be to those dualistic thinkers still among us, the power of human institutional behavior rests on the greatly expanded scope for instinct provided by the evolution of instrumentality. Furthermore, there is reason to believe that some of the phyletic discontinuities of the brain that make possible this discontinuity of effect are themselves continuity phenomena at the level of neurochemistry.

FUNCTIONAL HALLUCINATION

As students of hominid origins are well aware, we are confronted with profound behavioral discrepancies between humans and apes, and yet geneticists tell us that these species are so similar that humans and chimpanzees have 99 percent of their protein structure in common (King and Wilson 1975). Morphologically, their brains are overwhelmingly similar in gross anatomy and composed of homologous organs. Cognitively, close correspondences, which would have been unthinkable ten years ago, are now generally acknowledged. On the basis of psychological and behavioral comparisons with other primates, it is possible to isolate psychobehavioral discontinuities that are probably of importance in the emergence of the family Hominidae. As argued here, among the most important features of this *psychological adaptive complex* are control of constructional activities by an image of the final product, a capacity for social induction of conceptual networks, an incorporation of the branchiomeric musculature into volitional and instrumental control, the development of cerebral laterality for both hand and vocal tract, the possibility of a new auditory sequence analyzer, and an altered threshold in the reward value of consummatory fear and aggression. Although seemingly disparate, these various changes may be more closely connected to each other than each, considered separately, would suggest. As already noted, the biological similarities and behavioral discrepancies between man and ape require very small genetic changes with very extensive effects on neural function. A small change in a neurochemical modulation system could produce such ramifying psychic divergences, because such modulation systems do in fact produce profound differences of state within an individual brain from one time to the next. The incorporation of a limbic system effector into the instrumental modality, the close relationship between language and limbic system functions such as growth, memory and attention, the functional hallucination of human constructional activity, the subjective truth value of words, the self-reinforcing aspect of the human agonistic system, the further differentiation of facial expression, and the direct coupling of neocortical and limbic processes in linguo-conceptual induction—these all point in the same direction: toward minor evolutionary changes in those long neurochemical tracts that arise in the most phylogenetically primitive parts of the primate core brain but modulate the state-specific operation of its most phylogenetically advanced structures (Jouvet 1972; Moruzzi 1972; Cooper et al. 1978)

Historically, the critical differences between humans and apes were sought in the neocortex because people wanted to find them there. Yet, there is already a good deal of evidence that schizophrenia, the action of hallucinogenic drugs, the altered reward values of self-stimulation, dreaming sleep, the hippocampal–ACTH axis, and affective states are all neurochemically mediated phenomena, even though the precise mechanisms are still widely controversial (Cooper et al. 1978; Snyder 1974; Pradhan 1975; Warburton 1975; Morgenson and Phillips 1978; Phillips and Morgenson 1978). Although workers in these various fields have recognized the empirical connection among these phenomena, an evolutionary perspective indicates that they are not simply connected in a loose sense. Rather, they are all aspects of a functional change intrinsically related to the hominid adaptation: a neurochemical gate for functional hallucination. If this line of reasoning is correct, then the major discontinuity phenomena delineated behaviorally as definitive of the hominid lineage are in fact continuity phenomena at the level of neurochemistry. The discontinuity is one of effect, not process; and unless contemporary biological understanding is fundamentally misconstrued, the emergence of the Hominidae is in fact a problem of "normal science" in neurochemistry as presently practiced. This conclusion, so divergent from the traditional position, in fact accords well with the anthropologist's intuition. To its credit, evolutionary anthropology has always assumed that a certain degree of madness was definitive of the human species. Disputes about the human status of the craftsmen are endless in discussions of the two–million–year history of tools. Yet in the Upper Pleistocene when cairns of bear skulls were built in the depths of caves, when skeletons were painted with red ochre and adorned with flowers, as revealed by associated pollen grains—no one doubts that the threshold of humanity had been crossed.

NOTES TO CHAPTER 7

1. See chap. 4.

2. See Gesell (1940) for the developmental changes of body image in children as revealed by drawings.

3. See also the observational learning and sign language studies (chap. 6). Some cooperative tool use is reported for monkeys (Beck 1973).

4. I have discussed the multimale troop as hominid precursor in more detail elsewhere (Reynolds 1976) and have summarized male care-giving.

5. Chodorow (1974) deals with the conflict between male social status and the "infantile dependency" component of male-female relations. The parent-child role dichotomy is similar to the theory of transactional analysis.

6. Even when females or young participate, there may be special features of the male intergroup signals, as in howler monkeys (Baldwin and Baldwin 1976). In other cases, the males are disproportionately represented in aggressive display or behavior, even though other kinds of behavior occur and other age and sex categories participate (Deag 1973; Lindburg 1971). The prosimian, *Lemur catta,* is exceptional in that females lead intergroup encounters (Budnitz and Dainis 1975). See Gautier-Hion (1975) and Leutenegger and Kelly (1977) for reviews of anthropoid primates. See also Bernstein (1964*a,* 1966), Hall (1960), Modahl and Eaton (1977), and references in chap. 5.

REFERENCES

BALDWIN, J. D., and BALDWIN, J. I. 1976. Vocalizations of howler monkeys (*Alouatta palliata*) in Southwestern Panama. *Folia Primatol.* 26:81–108.

BECK, BENJAMIN B. 1973. Cooperative tool use by captive hamadryas baboons. *Science* 182:594–597.

BERNSTEIN, IRWIN S. 1964a. Role of the dominant male rhesus monkey in response to external challenges to the group. *J. Comp. and Physiol. Psychol.* 57:404–406.

1964b. Group social patterns as influenced by removal and later reintroduction of the dominant male rhesus. *Psychological Reports* 14:3–10.

1966. An investigation of the organization of pigtail monkey groups through the use of challenges. *Primates* 7:471–480.

BERNSTEIN, IRWIN S.; GORDON, T. P.; and ROSE, R. 1974. Aggression and social controls in rhesus monkey *(Macaca mulatta)* groups revealed in group formation studies. *Folia Primatol.* 21:81–107.

BUDNITZ, NORMAN, and DAINIS, KATHRYN 1975. *Lemur catta:* ecology and behavior. In *Lemur biology,* ed. I. Tattersall and R. W. Sussman, pp. 219–235. New York and London: Plenum.

BYGOTT, J. D. 1972. Cannibalism among wild chimpanzees. *Nature* (London) 238:410–411.

CHODOROW, NANCY 1974. Family structure and feminine personality. In *Woman, Culture, and Society,* ed. M. Z. Rosaldo and L. Lamphere, pp. 43–66. Stanford: Stanford U. Press.

COOPER, JACK R.; BLOOM, FLOYD E.; and ROTH, ROBERT H. 1978. *The biochemical basis of neuropharmacology.* 3rd ed. New York: Oxford U. Press.

CRAWFORD, MEREDITH P. 1937. The cooperative solving of problems by young chimpanzees. *Comp. Psychol. Monographs* 14, no. 2.

1941. The cooperative solving by chimpanzees of problems requiring serial responses to color cues. *J. Social Psychol.* 13:259–280.

CROOK, J. H. 1971. Sources of cooperation in animals and man. In *Man and beast: comparative social behavior,* ed. J. F. Eisenberg and W. S. Dillon, pp. 237–260. Washington, D.C.: Smithsonian Institution Press.

CROOK, JOHN HURRELL   1972. Sexual selection, dimorphism, and social organization in the primates. In *Sexual selection and the descent of man: 1871–1971*, ed. B. Campbell, pp. 231–281. Chicago: Aldine.

DAZEY, J.; KUYK, K.; OSWALD, M.; MARTENSON, J.; and ERWIN, J.   1974. Effects of group composition on agonistic behavior of captive pigtail macaques, *Macaca nemestrina. Amer. J. Physical Anthropol.* 46:73–76.

DEAG, JOHN M.   1973. Intergroup encounters in the wild Barbary macaque *Macaca sylvanus*. In *Comparative ecology and behavior of primates*, ed. R. P. Michael and J. H. Crook, pp. 315–373. New York and London: Academic Press.

DEAG, J. M., and CROOK, J. H.   1971. Social behavior and "agonistic buffering" in the wild barbary macaque *Macaca sylvana* L. *Folia Primatol.* 15:183–200.

EIBL-EIBESFELDT, I.   1971. *Love and hate.* Trans. Geoffrey Strachan. London: Methuen. 1st German ed., 1970.

FOX, ROBIN   1975. Primate kin and human kinship. In *Biosocial anthropology*, ed. Robin Fox, pp. 9–35. London: Malaby Press.

GAUTIER-HION, ANNIE   1975. Dimorphisme sexuel et organisation sociale chez les cercopithecines forestiers africains. *Mammalia* 39:365–374.

GESELL, A.   1940. *The child from one to five.* London: Methuen.

HALL, K. R. L.   1960. Social vigilance behaviour of the chacma baboon, *Papio ursinus. Behaviour* 16:261–284.

HANBY, J. P.; ROBERTSON, L. T.; and PHOENIX, C. H.   1971. The sexual behavior of a confined troop of Japanese macaques. *Folia Primatol.* 16:123–143.

HENDRICKS, D. C.; SEAY, B.; and BARNES, B.   1975. Effects of removal of dominant animals in a small group of *Macaca fascicularis. J. Gen. Psychol.* 92:157–168.

HERSHKOVITZ, PHILIP   1977. *Living New World primates (Platyrrhini), with an introduction to primates,* vol 1. Chicago: Chicago U. Press.

HOWELL, F. C.   1965. *Early Man.* New York: Time, Inc.

ISAAC, GLYNN   1975. Early hominids in action: a commentary on the contribution of archeology to understanding the fossil record in East Africa. *Yrbk. Physical Anthropology* 19:19–35.

1976. Stages of cultural elaboration in the Pleistocene: possible archaeological indicators of the development of language capabilities. *Annals N.Y. Acad. Science* 280:275–288.

1978. The food-sharing behavior of protohuman hominids. *Scientific American*, April, pp. 90–108.

JOUVET, M.   1972. The role of monoamines and acetylcholine-containing neurons in the regulation of the sleep-waking cycle. *Reviews of Physiology* 64:166–307. Berlin and New York: Springer Verlag.

KING, MARY-CLAIRE, and WILSON, A. C. 1975. Evolution at two levels in humans and chimpanzees. *Science* 188:107–116.

LAWICK-GOODALL, J. van 1975. The behaviour of the chimpanzee. In *Hominisation und verhalten*, ed. G. Kurth and I. Eibl-Eibesfeldt, pp. 74–136. Stuttgart: Gustav Fischer Verlag.

LEMON, W. B., and ALLEN, M. L. 1978. Continual sexual receptivity in the female chimpanzee *(Pan troglodytes)*. *Folia Primatol.* 30:80–88.

LEUTENEGGER, W., and KELLY, J. T. 1977. Relationship of sexual dimorphism in canine size and body size to social, behavioral, and ecological correlates in anthropoid primates. *Primates* 18:117–136.

LINDBURGH, D. 1971. The rhesus monkey in North India: an ecological and behavioral study. In *Primate behavior: developments in field and laboratory research*, ed. L. Rosenblum, vol. 2, pp. 1–106. New York: Academic Press.

LOY, J. 1971. Estrous behavior of free-ranging rhesus monkeys *(Macaca mulatta)*. *Primates* 12:1–31.

McDERMOTT, R. P.; GOSPODINOFF, K.; and ARON, JEFFREY 1978. Criteria for an ethnographically adequate description of concerted activities and their contexts. *Semiotica* 24:245–275.

MODAHL, K. B., and EATON, G. G. 1977. Display behaviour in a confined troop of Japanese macaques *(Macaca fuscata)*. *Animal Behaviour* 25:525–535.

MOGENSON, G. J., and PHILLIPS, A. G. 1978. Brain stimulation reward after 25 years. *Canadian J. Psychol.* 32:54–57.

MORUZZI, G. 1972. The sleep-waking cycle. *Reviews of Physiology* 64:1–165. Berlin and New York: Springer Verlag.

NEVILLE, MELVIN K. 1968b. A free-ranging rhesus monkey troop lacking adult males. *J. Mammalogy* 49:771–773.

NISHIDA, TOSHISADA 1968. The social group of wild chimpanzees in the Mahali Mountains. *Primates* 9:167–224.

NISSEN, H. W., and CRAWFORD, M. P. 1936. A preliminary study of food-sharing behavior in young chimpanzees. *J. Comp. Psychol.* 22:383–419.

OSWALD, M., and ERWIN, J. 1976. Control of intragroup aggression by male pigtail monkeys *(Macaca nemestrina)*. *Nature* (London) 262:686–688.

PHILLIPS, A. G., and MOGENSON, G. J. 1978. Brain-stimulation reward: current issues and future prospects. *Canadian J. Psychol.* 32:124–128. 32:124–128.

PRADHAN, S. N. 1975. Balances among central neurotransmitters in self-stimulation behavior. In *Neurotransmitter balances regulating behavior*, ed. E. F. Domino and J. M. Davis, pp. 75–97. Ann Arbor.

REYNOLDS, PETER C. 1976. The emergence of early hominid social organization. I. The attachment systems. *Yrbk. Physical Anthropology* 20:73–95.

REYNOLDS, VERNON, and REYNOLDS, FRANCES   1965. Chimpanzees of the Budongo forest. In *Primate behavior: field studies of monkeys and apes*, ed. Irven Devore, pp. 368–424. New York: Holt, Rinehart and Winston.

SACKETT, D. P.; OSWALD, M.; and ERWIN, M. J.   1975. Aggression among captive female pigtail monkeys in all-female and harem groups. *J. Biol. Psychol.* 17:17–20.

STEKLIS, H. D., and KING, G. E.   1978. The cranio-cervical killing bite. *J. Human Evolution* 7:567–581.

SNYDER, S. H.   1974. *Madness and the brain.* New York: McGraw-Hill.

TOKUDA, K., and JENSEN, G.   1968. The leader's role in controlling aggressive behavior in a monkey group. *Primates* 9:319–322.

TRIVERS, ROBERT L.   1972. Parental investment and sexual selection. In *Sexual selection and the descent of man: 1871–1971*, ed. B. Campbell, pp. 136–179. Chicago: Aldine.

WARBURTON, DAVID M.   1975. *Brain, behaviour and drugs: introduction to the neurochemistry of behavior.* London: John Wiley.

WICKLER, W.   1967. Socio-sexual signals and their intra-specific imitation among primates. In *Primate Ethology*, ed. D. Morris, pp. 69–147. Chicago: Aldine.

CONCLUSION

Is social inequality ordained by natural law? In raising Rousseau's question two centuries later, and in attempting to answer it by attention to the behavior of other primates, as he himself suggested, we are led to some conclusions that are strongly at variance with the conventional wisdom of anthropological theory. There *is* a natural component to social hierarchy, as conservatives have always maintained, but they will not be happy to learn that the prototype of exchange is grooming not aggression. There is also a natural component to the human family, as conservatives have likewise maintained, but nature does not point toward the nuclear family dominated by the Victorian father but to the matrilines and semipromiscuous mating of the multimale troop. There is a biological component to human kinship, as many anthropologists have argued, but the biological basis is not mating but attachment mechanisms. The opposite end of the theoretical spectrum fares no better. The social relationships of primates, far from being autonomous social facts, are created and sustained by innate motivational systems; and social attachment itself is largely derivative, ontogenetically and phylogenetically, from fear and aggression. The basic premise of socialization theory, that learning replaces instinct, has no support from the comparative study of primates, and the methodologies and conclusions that follow from this premise are unsupported as well. Contrary to the stratified view of human nature enshrined in anthropological theory, cultural integration requires instinct in order to work. Even worse, the concept of the instrumental modality of action challenges socialization theory on its home ground: on the march of technical progress itself. If the argument set forth in this book is correct, then the human species is characterized by the ability to integrate the socially generated programs of cooperative, instrumental action into affective mechanisms not substantially different from those of apes. If biological continuity lies in the mechanisms of control and integration of behavior, not in the morphology of behavior itself, then it follows that the most advanced technical programs of the human species—their cultural content notwithstanding—can be *homologous*, in a strict sense of the term, with the simplest instrumental acts of monkeys. Consequently, it is necessary to seriously consider the possibility that all primate social organizations, since they are to some extent created through common mechanisms of affect and instrumentality, are to some extent homologous too. In the light of these conclusions, the only course open to social theory is to stare biology square in the face and start again from the beginning.

GLOSSARY

_____

Terms used in the text are defined in this glossary if they are not likely to be found in the average dictionary or if they are used with altered or restricted meanings.

AFFECT   A (1) behavioral and (2) motivational system specialized for (3) the assessment of social consequences of action and (4) the assessment of the consequences of external events on oneself which is (5) mediated by the LIMBIC SYSTEM structures of the mammalian brain and is preferentially connected to (6) particular innate instrumental acts, (7) specific patterns of display, and (8) specific CONSUMMATORY BEHAVIORS.

AFFECTIVE-INSTRUMENTAL INTEGRATION   Behaviors that are produced by the mutual interaction of the AFFECTIVE and INSTRUMENTAL MODALITIES.

AFFECTIVE MODALITY   A functional system of the brain which assesses and anticipates the social consequences of action and the consequences of action on oneself. The AFFECTIVE MODALITY THEORY maintains that progressive instrumental action becomes incorporated into affectivity in the course of primate evolution.

AGGRESSION   Behavior which is appropriate to inflicting pain on another or the display behavior normally associated with it or behavior which is under the control of that motivation.

AGONISM   A generic term for the behavior patterns and motivations involved in FEAR, anger, escape, and AGGRESSION.

AGONISTIC DERIVATIVES   Motivational and behavioral systems hypothesized to derive from the differentiation of AGONISTIC behavior in the course of phylogeny, primarily by the development of new stable threshold levels within an existing spectrum.

_____

AMERICAN SIGN LANGUAGE or ASL or AMESLAN   The sign language used by the deaf in the United States. It is distinct from British sign language and from finger spelling.

AMYGDALA   See LIMBIC SYSTEM.

ANGULAR GYRUS   A GYRUS in the region of the inferior parietal lobe (*see* CORTEX) and the posterior portion of the middle temporal gyrus. The cortex in this geographical area is important in the neural control of language in man.

APHASIA   Loss or disability of language because of damage to the central (as opposed to peripheral) nervous system and which is not explicable simply as a more general sensory or motor impairment.

APPETITIVE BEHAVIOR   Actions that are instrumental to the opportunity to engage in a particular CONSUMMATORY ACT.

ASSOCIATION CORTEX   NEOCORTEX that receives its input from other cortical areas or from the intrinsic THALAMUS rather than from direct sensory pathways. It is theorized to play an important role in the evolution of primate intelligence.

ATHALAMIC   Regions of the NEOCORTEX which are believed to be lacking in direct projections from the THALAMUS, and, by implication, to receive their input from other neocortical regions.

ATTACHMENT   A prolonged social relationship between two or more individuals, characterized by mutual interaction and by evidence of affective arousal when the relationship is severed or threatened.

AUTONOMIC or VEGETATIVE or CRANIOSACRAL NERVOUS SYSTEM   That portion of the nervous system which controls SMOOTH MUSCLE, cardiac muscle, and glands.

CONSUMMATORY ACT   The action that normally terminates a motivated sequence of actions, as when feeding terminates food-getting behavior or biting terminates aggression. Contrasts to APPETITIVE BEHAVIOR.

APPROACH-AVOIDANCE GRADIENT   Behavior characterized by oscillations between approach and avoidance to a source of stimulation, sometimes leading to immobility.

BASAL GANGLIA   A generic term for certain NUCLEI in the forebrain, including corpus STRIATUM and putamen.

BEHAVIOR   Observable movement in an animal produced by neuromuscular activity.

BEHAVIOR PATTERN   A generic term for some behavioral unit that is readily distinguishable and useful in behavioral description.

BRAIN   The neural tissue that is found in the skull. During development of the embryo in vertebrates, the neural tube differentiates into three *primary vesicles*. The anterior vesicle gives rise to neural structures which are collectively termed the FOREBRAIN or *PROSENCEPHALON* and among these are the NEOCORTEX, HIPPOCAMPUS, BASAL GANGLIA, THALAMUS, and the HYPOTHALAMUS. The middle vesicle gives rise to the MIDBRAIN or *MESENCEPHALON*, while the posterior vesicle gives rise to the *HINDBRAIN*, including such important structures as the CEREBELLUM and medulla oblongata.

BRAINSTEM   That tubelike portion of the brain which connects the forebrain to the spinal cord. It includes such anatomically distinct parts as the midbrain and medulla oblongata (*see* BRAIN).

BRANCHIOMERIC   Pertaining to the *branchiomeres* or segments of the *splanchnic* (visceral) mesoderm from which the *branchial arches* (supporting structures for gills) develop embryologically in vertebrates. Portions of the vertebrate skull which are derived from the branchial arches are also referred to as the *viscerocranium*.

CEREBELLUM   A large structure in the posterior portion of the brain that is involved in the control of movement.

CLASSICAL ETHOLOGY   A theory of animal behavior developed by Lorenz, Tinbergen, and their coworkers, which emphasized the theoretical constructs of FIXED-ACTION PATTERN, action-specific energy, and INNATE RELEASING MECHANISM.

CONCEPT   *See* CONCEPTUALIZATION; DISJUNCTIVE CONCEPT; CONJUNCTIVE CONCEPT; INSTRUMENTAL CONCEPT; and CONCEPT EXEMPLAR.

CONCEPT EXEMPLAR   The current perceptual information believed to exemplify a particular concept.

CONCEPTUALIZATION   The process of encoding information such that it is amenable to mental manipulation through inductive and deductive logic and also the transformations of stored information produced by such manipulations.

CONJUNCTIVE CONCEPT   A concept in which all the exemplars share common perceptual recognition criteria, such as "square."

CONSORT PAIR   The tendency of mating couples to form a subgroup that acts as a unit vis-à-vis other members of the troop.

CONSTRUCTIONAL SYSTEM   A functional system in the brain involved in the conceptualization of relationships among objects, in the motor control of behaviors which create alterations of existing object relationships and modifications of particular objects, and in the mental manipulation of representations of objects encoded in iconic form.

COOPERATIVE BEHAVIOR   Def. 1. Behavior that is INTENDED to be INSTRUMENTAL to the inferred goals of another's action. Def 2. Behavior that is effective in producing a common result through processes of social coordination.

COOPERATIVE HUNTING   Hunting in which multiple individuals coordinate their activity relative to the common goal of catching a particular prey target.

CORE BRAIN   A generic term for those structures lying near the medial surface of the brain, especially those involved in VISCEROAUTONOMIC activity.

CORTEX (CEREBRAL CORTEX) OR PALLIUM   The mantle of neural tissue that covers the cerebral hemisphere portion of the forebrain. It includes both NEOCORTEX and other cortical regions which are distinguished from one another by the number of distinct cell layers when examined under a microscope. The cortex is divided into four main geographical regions or *lobes* for each cerebral hemisphere: an anterior portion or *frontal lobe*, an upper central portion or *parietal lobe*, a lower central portion or *temporal lobe*, and a posterior portion or *occipital lobe*. The neocortical component expands dramatically in primate evolution and constitutes the bulk of the brain by volume. The surface of the cortex is folded into GYRI.

CRANIAL NERVES   The twelve pairs of nerves that are connected to the brain (rather than to the spinal cord): Olfactory (I), Optic (II), Oculomotor (III), Trochlear (IV), Trigeminal (V), Abducens (VI), Facial (VII), Vestibulocochlear (VIII), Glossopharyngeal (IX), Vagus (X), Accessory (XI), and Hypoglossal (XII).

DIFFUSE TOUCH   Touch sensation that is not primarily specialized for the transmission of precise spatiotemporal localization.

DISJUNCTIVE CONCEPT   A concept in which all the exemplars do not share common perceptual recognition criteria, such as a "strike" in baseball.

EPICRITIC   Cutaneous (skin) nerve fibers that are specialized for the transmission of fine distinctions in stimulation. Pribram has generalized the term to neural systems in which precise spatiotemporal localization is preserved through encoding and transmission.

ETHOLOGY   A generic term for theories of behavior which emphasize genetic, evolutionary, and innate motivational mechanisms in the causation of behavior.

EXCHANGE SYSTEMS   The social distribution of rewards through volitional and intentional action or the resulting networks of reward distribution.

EXTEROCEPTOR   A sense organ that relays information from the body surface and which normally responds to external stimuli.

EXTRAPYRAMIDAL MOTOR SYSTEM   A generic term for those neural structures involved in the regulation of posture and gross movement. It includes the STRIATUM, red nucleus, substantia nigra, and related connections to the CEREBELLUM and the CORTEX. The term *extrapyramidal* contrasts to PYRAMIDAL TRACT.

FEAR   Behavior that is appropriate to escape or the display behavior normally associated with it or behavior under the control of that motivation.

FINE TUNING   The control of a motivational system so that it is capable of a wide range of response alternatives and degrees of intensity.

FIXED-ACTION PATTERN   A behavioral unit, relatively stereotyped in form, which generally develops in animals raised in social isolation and is also characteristic of normal members of the species.

FOREBRAIN   See BRAIN.

FRONTAL CORTEX   See CORTEX.

FUNCTIONAL HALLUCINATION   The hypothesis that evolutionary changes in the brain mechanisms underlying hallucinations are fundamental to the psychological changes that distinguish hominids from other primates.

GREAT CHAIN OF BEING   Lovejoy's term for the belief that living things are arranged in a hierarchy from the more perfect to the less perfect.

GYRUS   A ridge formed by the folding of the external surface of the NEOCORTEX. Gyri are used as geographical landmarks but are not functional units in themselves.

HIPPOCAMPUS   See LIMBIC SYSTEM.

HISTORICISM   The belief that history is not simply a record of past human activity but is a natural process subject to discoverable laws.

HOMOLOGY   Before Darwin, the classification of the organs of different species by similar position or relationship relative to other organs. After Darwin, those organs of different species that are descended from a common ancestor.

HYPOTHALAMUS   A part of the forebrain (see BRAIN) which regulates neuroendocrine interactions, such as water balance, body temperature, sexuality, hunger, and growth, and which serves as the highest integrative center for the AUTONOMIC NERVOUS SYSTEM.

ICONIC DRIVING   The hypothesis that hominid CONSTRUCTIONAL activity requires the control of behavior by a simulated image of future perceptual information, or, more generally, the use of iconic shape information to alter the shape of a body part.

INFEROTEMPORAL CORTEX   The neocortex of the inferior surface of the temporal lobe (see CORTEX). Its removal in monkeys causes a deficit in learned visual discrimination tasks.

INHIBITED ESCAPE   A theoretical construct that explains certain aspects of primate affiliation, such as smiling, as an elevation of the threshold of escape behavior such that the reward value of actual escape occurs in the absence of overt escape behavior.

INNATE-LEARNING INTERLOCKING   A behavioral system whose normal development requires the interaction between innate and environmental information.

INNATE RELEASING MECHANISM An inborn perceptual mechanism that activates a particular innate behavior when a particular innately encoded stimulus configuration is perceived in the external world.

INSTITUTION A social group that produces a collective product or effect through COOPERATIVE, INTENTIONAL, INSTRUMENTAL action.

INSTRUMENTAL BEHAVIOR An action that is instrumental to achieving a particular effect, or, as used here, behavior that is under the control of the INSTRUMENTAL MODALITY.

INSTRUMENTAL CHANNEL THEORY OF LANGUAGE The theory that language is a central system that uses whatever mechanisms the instrumental modality makes available to it, such that the behavioral complexity of language will vary with phyletic differences in the complexity of instrumental action.

INSTRUMENTAL CONCEPT The class of behaviors or techniques conceived of as instrumental to the same effect.

INSTRUMENTAL MODALITY A functional system of the brain which uses the SPECIAL SENSES and CONCEPTUALIZATION to assess and anticipate the consequences of action. The *instrumental modality theory* maintains that progressively more distal body parts and more complex programs become integrated into the modality in the course of primate evolution, culminating in the integration of cooperative object-using programs in hominids.

INTENTIONAL ACTION Action in which the consequences are anticipated through CONCEPTUALIZATION prior to execution.

INTRACORTICAL Connections that go directly from one region of neocortex to another.

LATERALIZATION The tendency of one side of a bilateral brain structure to assume control of a particular function irrespective of the side of the body which performs the activity.

LIMBIC SYSTEM An anatomically diverse collection of structures in the mammalian forebrain (*see* BRAIN) involved in the regulation of affect, learning, memory, and motivation. It includes the hippocampus and dentate gyri (which are underneath the neocortical mantle on the medial surface of the cerebral hemispheres), the BASAL GANGLIA structure called the amygdala, and the structures in the septal area connected to the sensory pathways for smell.

MIDBRAIN *See* BRAIN.

MODALITIES OF RECOMBINATION The hypothesis that there are distinct functional systems within the brain, sustained by distinct motivations and central states, which are specialized for the merging of innate and acquired information into composite behavioral products in the course of maturation.

MORPHOLOGY OF BEHAVIOR The classification of behaviors by their similarity in appearance.

MULTIMALE TROOP   A type of primate social organization characterized by the presence of multiple breeding adult males.

NEOCORTEX OR NEOPALLIUM   That portion of the CORTEX which is characterized by six distinct cell layers when examined under a microscope. When viewed with the naked eye, the external surface of the neocortex is folded into ridges called GYRI and ravines called SULCI or FISSURES. Many sulci and gyri are named, such as the ANGULAR GYRUS and SU-PRAMARGINAL GYRUS.

NEUROCHEMICAL GATE   The hypothesis that psychological changes that occur in the hominid lineage, particularly those relating to FUNCTIONAL HALLUCINATION, are owing primarily to altered functioning among existing brain structures through alterations in the neurochemical environment of those structures.

NEUROCHEMICAL MODULATION   Alteration in the physiological activity of brain structures through changes in the chemical environment of the structure or of its specialized receptor organs.

NOBLE SAVAGE   The belief that man is innocent in a state of nature and becomes corrupted by civilization.

NUCLEUS (Pl. NUCLEI)   A distinguishable cluster of nerve cells in the central nervous system.

OVERLEARNING   Tasks that have become habitual may be preserved after removal of the parts of the neocortex that are necessary for the task in other circumstances, and such tasks are then called overlearned.

PARIETAL CORTEX   See CORTEX.

PART BEHAVIOR   A motor act that normally occurs as a component of a more complex behavior and not in isolation.

PRECONSUMMATORY REWARD   A theoretical concept that explains the evolution of certain behavioral systems, such as play or smiling, as a phylogenetic shift in the threshold of a CONSUMMATORY ACT, such that the reward value appropriate to the consummatory act is achieved by the behavioral act that ordinarily just precedes it.

PRESTATIONAL BEHAVIOR   Volitional behavior that is directed to another and in which the recipient has the option of accepting or refusing.

PROTOCRITIC   Pribram's term for neural systems in which precise spatiotemporal information is lost during encoding and transmission. The term is derivative of *protopathic sensibility* which is the poorly localized responsiveness to pain and temperature which may remain after nerve damage.

PULVINAR   See THALAMUS.

PYRAMIDAL OR CORTICOSPINAL TRACT   A tract of nerve fibers that arises in the CORTEX and descends directly to motor neurons in the spinal cord. It is theorized to be involved in the fine motor control of the extremities.

RANKING SYSTEM A social group whose members can be ranked vis-à-vis one another in terms of the ratio of wins to losses in aggressive encounters.

RETICULAR FORMATION A netlike collection of interconnected nerve cells in the central core of the BRAINSTEM, extending from the midbrain (see BRAIN) caudally to the boundary with the spinal cord. Some of its constituent NUCLEI are involved in arousal to significant stimuli and in altered states of the whole brain.

SELF CONCEPT The conceptualization of one's own body and its relationship to other entities.

SKELETAL OR SOMATIC MUSCULATURE STRIPED MUSCLES attached to bones.

SMOOTH MUSCULATURE Nonstriate muscle not normally subject to voluntary control. It is associated with organs under the control of the AUTONOMIC NERVOUS SYSTEM, such as glands, viscera, and blood vessels. (See STRIPED MUSCULATURE.)

SOMATIC MUSCULATURE See SKELETAL MUSCULATURE.

SOMATIC NERVOUS SYSTEM That portion of the nervous system involved in the transmission of information to and from the SKELETAL MUSCLES, bones, joints, ligaments, EXTEROCEPTORS and other nonvisceral structures.

SPECIAL SENSES The senses of vision, touch, hearing, taste, and smell.

SPECIES-SPECIFIC Behaviors characteristic of a particular species, especially those that distinguish it from closely related species.

SPEECH HEMISPHERE The cerebral hemisphere in man that is LATERALIZED for the control of speech, whichever hemisphere that may be in a particular individual, or alternatively, the normal speech hemisphere, namely the left.

SPINOTHALAMIC TRACT A tract of nerve fibers which carries pain and temperature information from the body via the spinal cord and terminates in the THALAMUS.

STRIATUM OR CORPUS STRIATUM A portion of the BASAL GANGLIA, including the caudate nucleus and lentiform nucleus, involved in the regulation of the movements of locomotion, posture, and facial expression. The striatum is a major component of the EXTRAPYRAMIDAL MOTOR SYSTEM.

STRIPED OR STRIATE MUSCULATURE Muscles whose fibers exhibit parallel bands when examined under a microscope. Striate muscles occur in the SKELETAL MUSCULATURE where they may be under voluntary control and in the cardiac musculature where they are not.

SUPRAMARGINAL GYRUS A GYRUS in the region of the inferior parietal lobe and the superior temporal gyrus (see CORTEX; NEOCORTEX).

SYSTEMIC-FORMATIVE CODE   A communicative system in which elementary
   articulatory units are systematically and simultaneously combined to produce
   perceptually discrete units, and these units in turn are combined to form
   semantically distinct units. Both speech and deaf sign languages are members
   of this class.

TEMPORAL CORTEX   *See* CORTEX.

THALAMOCORTICAL   Pertaining to the interaction between THALAMUS
   and CORTEX.

THALAMUS   A region of the forebrain from which the CORTEX receives its
   input. The thalamus is divided into many nuclei, each of which sends specific
   kinds of information to specific regions of the cortex. Thalamic nuclei in
   turn are regarded as *extrinsic* nuclei if they receive their input from outside
   the thalamus and *intrinsic* nuclei if they receive their input from other thalamic
   nuclei. The medial geniculate body is an example of an extrinsic nucleus
   since it receives auditory input from the eighth nerve and sends information
   to the auditory cortex. The pulvinar is an intrinsic thalamic nucleus: it is not
   directly related to sensory pathways and projects to the ASSOCIATION
   CORTEX.

TRIGEMINAL NERVE   The fifth CRANIAL NERVE. It controls the muscles
   used in chewing and relays sensory information from the teeth, face, mouth,
   and nasal cavity. It also has important central components in the trigeminal
   nuclei in the brain.

VICTORIAN BRAIN   The belief that the brain is primarily organized in terms
   of phylogenetically primitive and advanced structures, that the former are
   subordinated to the latter, and that AFFECT is primarily primitive and
   individualistic.

VISCERAL   Pertaining to the internal organs and particularly to those differentiated regions of the embryo which give rise to the viscera and related organs.

VISCEROAUTONOMIC   The portion of the nervous system concerned with the
   regulation and control of cardiac muscle, SMOOTH MUSCLE, and glands
   or with those STRIPED MUSCLE systems derived phylogenetically from
   VISCERAL structures.

VOLITIONAL BEHAVIOR   Behavior that can be inhibited through CONCEP-
   TUALIZATION.

INDEX

Affective expression, 35–36, 67, 82, 89–140, 248–249
Affective modality, 80–83, 89–140, 217, 232–233
Affective-instrumental integration, 82–83, 128–129, 151, 217, 236, 249
Aggression. *See* Agonism
Agonism, 90, 94–96, 100, 106–129, 195–198
Agonistic buffering, 152–153, 245–246
Anthropology, history of, 1–61
Ape language training, 147–149, 158, 162–163, 211, 214–219, 231–232
Aphasia, 221–223
Approach-avoidance behavior, 97–98, 106, 109, 112–113, 120, 122, 172–173
Attachment, 98–99, 110–115, 121, 127–129, 176–186, 243–245

Bird song, 22–23, 31–33, 233
Buffon, Count, 4–5, 11, 39–40

Comte, Auguste, 33, 69–70
Conceptualization, 63–74, 73, 83, 141–170, 215, 219, 229, 232–234
Constructional ability, 156–163
Culture, 14–18, 33, 164, 210–211, 228, 247–248

Darwin, Charles, 4–5, 10–13, 35–36, 38
Dreaming sleep, 74–79, 234
Durkheim, Émile, 19–21, 36–37, 71

Exchange, 171–208, 245–246; of food, 194–198, 243–244. *See also* Prestational processes
Eye-spot, 98, 102–104, 120, 244–245, 248

Fear. *See* Agonism
Fine tuning of agonism, 106–110, 112–120
Fixed-action pattern, 64, 67–69, 72–74, 78, 92–93
Freud, Sigmund, 19, 36, 64, 76, 211, 224, 231, 248

Great Chain of Being, 1–61
Grooming, 82, 171–208, 243–244

Homology, 4–5, 24–27, 64–65, 102
Hypothalamus, 67–68, 108, 117

Iconic driving, 160–163, 211, 249
Instinct, 18–19, 31–33, 35–37, 63–88, 100–101, 248–249
Institution, 70–74, 83, 242, 246, 248–249